Reproduction in mammals

Book 4: Reproductive fitness

SECOND EDITION

Reproduction in mammals

BOOK 4

Reproductive fitness

EDITED BY C. R. AUSTIN

Formerly Fellow of Fitzwilliam College
Emeritus Charles Darwin Professor of Animal Embryology
University of Cambridge

AND R. V. SHORT, FRS

Professor of Reproductive Biology
Monash University, Melbourne, Australia

DRAWINGS BY JOHN R. FULLER

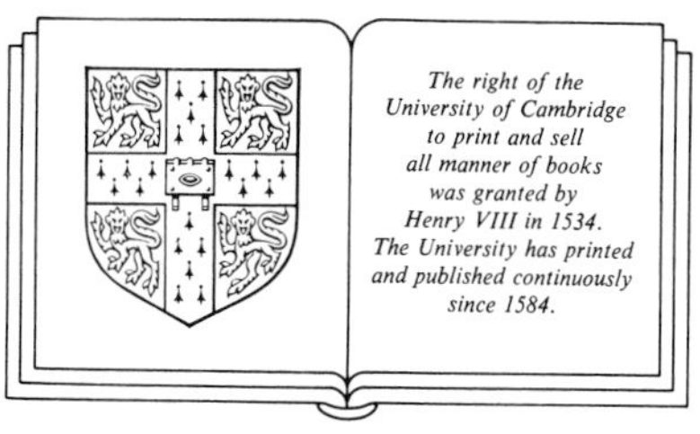

Cambridge University Press

Cambridge

New York Port Chester

Melbourne Sydney

Published by the Press Syndicate of the University of Cambridge
The Pitt Building, Trumpington Street, Cambridge CB2 1RP
40 West 20th Street, New York, NY 10011, USA
10 Stamford Road, Oakleigh, Melbourne 3166, Australia

First published as *Reproductive Patterns* 1972
Reprinted 1973
Second edition 1985
Reprinted 1989

Printed in Great Britain by the University Press, Cambridge

Library of Congress catalogue card number: 81-18060

British Library cataloguing in publication data

Reproduction in mammals. – 2nd ed
Bk. 4: Reproductive fitness

1. Mammals – Reproduction
I. Austin, C. R. II. Short, R. V.
599.01′6 QP251

ISBN 0 521 26649 1 hard covers
ISBN 0 521 31984 6 paperback
(First edition:
ISBN 0 521 08578 0 hard covers
ISBN 0 521 09616 2 paperback)

UP

CONTENTS

CONTRIBUTORS TO BOOK 4

C. E. Adams
ARC Institute of Animal Physiology
Animal Research Station
307 Huntingdon Road
Cambridge CB2 4AT, UK

N. J. Alexander
Reproductive Physiology
Oregon Regional Primate Research Center
Beaverton, Oregon 97006, USA

D. J. Anderson
Division of Immunogenetics
Sidney Farber Cancer Institute
Harvard Medical School
Boston, Massachusetts 02115, USA

B. K. Follet, FRS
ARC Research Group on Photoperiodism & Reproduction
Department of Zoology
The University
Bristol BS8 1UG, UK

E. B. Keverne
Department of Anatomy
Cambridge University
Downing Street
Cambridge CB2 3DY, UK

R. B. Land
ARC Animal Breeding Research Organization
West Mains Road
Edinburgh EH9 3JQ, UK

R. M. May, FRS
Biology Department
Princeton University
Princeton, New Jersey 08544, USA

D. I. Rubenstein
Biology Department
Princeton University
Princeton, New Jersey 08544, USA

R. V. Short, FRS
Department of Physiology
Monash University
Clayton, Victoria 3168, Australia

PREFACE TO THE SECOND EDITION

In this, our Second Edition of *Reproduction in Mammals*, we are responding to numerous requests for a more up-to-date and rather more detailed treatment of the subject. The First Edition was accorded an excellent reception, but the Books 1 to 5 were written 12 years ago and inevitably there have been advances on many fronts since then. As before, the manner of presentation is intended to make the subject matter interesting to read and readily comprehensible to undergraduates in the biological sciences, and yet with sufficient depth to provide a valued source of information to graduates engaged in both teaching and research. Our authors have been selected from among the best known in their respective fields.

Book 4 pays particular attention to genetic, environmental, behavioural and immunological mechanisms that can contribute to an animal's overall reproductive fitness, through which natural selection must ultimately operate. Mammals have developed a spectacular array of reproductive adaptations in order to meet the demands of a wide variety of different environmental settings, from ocean depths to arid deserts and snow-capped mountain peaks. The reproductive system has had to obey certain design constraints imposed by the animal's body size: small species with short lifespans can reproduce extremely rapidly, whereas the long-lived large animals must reproduce slowly. This fact has had profound repercussions on the development of the entire reproductive system, and is reflected also in the phenomenon of reproductive ageing.

From the Preface to the First Edition

Reproduction in Mammals is intended to meet the needs of undergraduates reading Zoology, Biology, Physiology, Medicine, Veterinary Science and Agriculture, and as a source of information for advanced students and research workers. It is published as a series of eight small textbooks dealing with all major aspects of mammalian reproduction. Each of the component books is designed to cover independently fairly distinct subdivisions of the subject, so that readers can select texts relevant to their particular interests and needs, if reluctant to purchase the whole work. The contents list of all the books are set out on the next page.

BOOKS IN THE FIRST EDITION

Book 1. Germ cells and fertilization (1972)
Primordial germ cells *T. G. Baker*
Oogenesis and ovulation *T. G. Baker*
Spermatogenesis and the spermatozoa *V. Monesi*
Cycles and seasons *R. M. F. S. Sadleir*
Fertilization *C. R. Austin*
Book 2. Embryonic and fetal development (1972)
The embryo *A. McLaren*
Sex determination and differentiation *R. V. Short*
The fetus and birth *G. C. Liggins*
Manipulation of development *R. L. Gardner*
Pregnancy losses and birth defects *C. R. Austin*
Book 3. Hormones in reproduction (1972)
Reproductive hormones *D. T. Baird*
The hypothalamus *B. A. Cross*
Role of hormones in sex cycles *R. V. Short*
Roles of hormones in pregnancy *R. B. Heap*
Lactation and its hormonal control *A. T. Cowie*
Book 4. Reproductive patterns (1972)
Species differences *R. V. Short*
Behavioural patterns *J. Herbert*
Environmental effects *R. M. F. S. Sadleir*
Immunological influences *R. G. Edwards*
Ageing and reproduction *C. E. Adams*
Book 5. Artificial control of reproduction (1972)
Increasing reproductive potential in farm animals *C. Polge*
Limiting human reproductive potential *D. M. Potts*
Chemical methods of male contraception *H. Jackson*
Control of human development *R. G. Edwards*
Reproduction and human society *R. V. Short*
The ethics of manipulating reproduction in man *C. R. Austin*
Book 6. The evolution of reproduction (1976)
The development of sexual reproduction *S. Ohno*
Evolution of viviparity in mammals *G. B. Sharman*
Selection for reproductive success *P. A. Jewell*
The origin of species *R. V. Short*
Specialization of gametes *C. R. Austin*
Book 7. Mechanisms of hormone action (1979)
Releasing hormones *H. M. Fraser*
Pituitary and placental hormones. *J. Dorrington*
Prostaglandins *J. R. G. Challis*
The androgens *W. I. P. Mainwaring*
The oestrogens *E. V. Jensen*
Progesterone *R. B. Heap and A. P. F. Flint*
Book 8. Human sexuality (1980)
The origins of human sexuality *R. V. Short*
Human sexual behaviour *J. Bancroft*
Variant forms of human sexual behaviour *R. Green*
Patterns of sexual behaviour in contemporary society *M. Schofield*
Constraints on sexual behaviour *C. R. Austin*
A perennial morality *G. R. Dunstan*

BOOKS IN THE SECOND EDITION

Book 1. Germ cells and fertilization (1982)
Primordial germ cells and the regulation of meiosis *A. G. Byskov*
Oogenesis and ovulation *T. G. Baker*
The egg *C. R. Austin*
Spermatogenesis and spermatozoa *B. P. Setchell*
Sperm and egg transport *M. J. K. Harper*
Fertilization *J. M. Bedford*
Book 2. Embryonic and fetal development (1982)
The embryo *A. McLaren*
Implantation and placentation *M. B. Renfree*
Sex determination and differentiation *R. V. Short*
The fetus and birth *G. C. Liggins*
Pregnancy losses and birth defects *P. A. Jacobs*
Manipulation of development *R. L. Gardner*
Book 3. Hormonal control of reproduction (1984)
The hypothalamus and anterior pituitary gland *F. J. Karsch*
The posterior pituitary *D. W. Lincoln*
The pineal gland *G. A. Lincoln*
The testis *D. M. de Kretser*
The ovary *D. T. Baird*
Oestrous and menstrual cycles *R. V. Short*
Pregnancy *R. B. Heap and A. P. F. Flint*
Lactation *A. T. Cowie*
Book 4. Reproductive fitness (1984)
Reproductive strategies *R. M. May and D. I. Rubenstein*
Species differences in reproductive mechanisms *R. V. Short*
Genetics and reproduction *R. B. Land*
The environment and reproduction *B. K. Follett*
Reproductive behaviour *E. B. Keverne*
Immunological factors in reproductive fitness *N. J. Alexander and D. J. Anderson*
Reproductive senescence *C. E. Adams*
Book 5. Manipulating reproduction (in Press)
Increasing productivity in farm animals *K. J. Betteridge*
Today's and tomorrow's contraceptives *R. V. Short*
Contraceptive needs of the developing world *D. M. Potts*
Benefits and risks of contraception *M. P. Vessey*
Alleviating human infertility *J. Cohen, C. B. Fehilly and R. G. Edwards*
Some reproductive options *A. McLaren*
Barriers to population control *C. R. Austin*

1

Reproductive strategies

ROBERT M. MAY AND DANIEL I. RUBENSTEIN

Natural populations of animals exhibit a bewildering variety of dynamic behaviour. For some species local abundance remains roughly unchanging, year after year. For example, in *The Natural History of Selborne* (arguably the first book on ecology, published in 1789), Gilbert White observed that the number of swifts flying around the church tower was approximately constant, at eight pairs, every year; the same number of swifts are to be found in Selborne in the summer today. The abundance of other species waxes and wanes (often by factors in excess of 10000) in well-defined cycles: such are the 4-year cycles in the numbers of mice, voles and lemmings in most northerly regions, the 10- to 11-year cycles in abundance of snowshoe hares and lynx and other predators in Canada, and the cycles in many insect pest species in temperate forests, with periods ranging from 5 to 12 years. Yet other natural populations exhibit irregular fluctuations, with episodes of outbreak or rarity often keyed to the weather: examples are the African desert locust, or the wasps on Gilbert White's fruit trees ('in 1781 we had none; in 1783 there were myriads'). Fig. 1.1 shows the variety of dynamic patterns exhibited by four vertebrate populations in Wytham Wood in England.

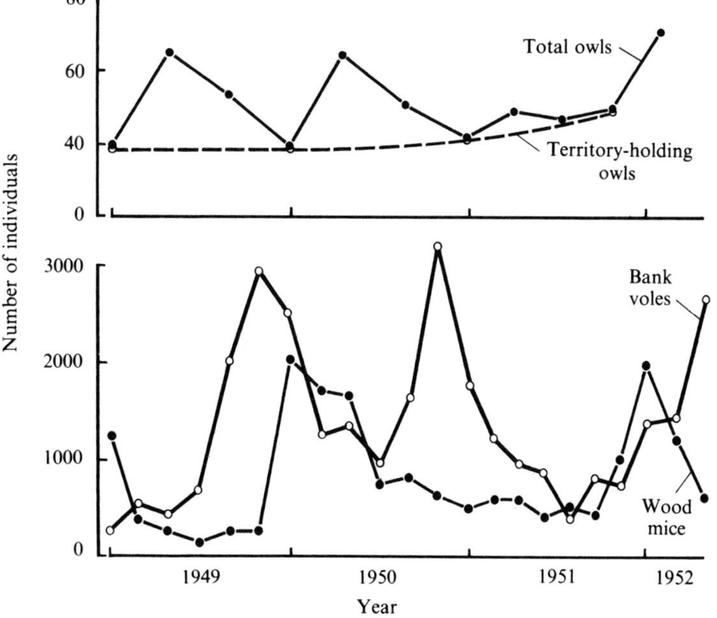

Fig. 1.1. Fluctuations in the numbers of total owls, territory-holding owls, bank voles and wood mice in Wytham Wood, 1949–52. (From T. R. E. Southwood. Bioeconomic strategies of population parameters. In *Theoretical Ecology*, pp. 26–68. Ed. R. M. May. Blackwell Scientific Publications; Oxford (1976).)

From Darwin's time to our own, much research has been directed towards codifying these patterns, and trying to understand them. In general, the overall dynamic behaviour of an animal population will depend on the character of the birth and death processes within it; these processes are forged – in an evolutionary 'furnace' – by the interactions between the population and its physical and biological environment. Thus, the reproductive biology of a mammalian species is ultimately entwined both with its evolutionary biology and with its population dynamics.

One crude generalization about the relation between a species' life-history strategy and its physical and biological setting invokes the deliberately oversimplified concept of r selection and K selection. The general ideas here were formulated by Darwin, Schmalhausen, Simpson, Stebbins and others, but it was MacArthur and Wilson who coined the phrase 'r and K selection', derived from the conventional parameters in the logistic equation, $\mathrm{d}N/\mathrm{d}t = rN(1-N/K)$. This equation describes the familiar sigmoid curve of ultimately bounded population growth: at low population densities there is essentially pure exponential growth, at the rate r; at high densities the population tends to stabilize around a value K which is set by some 'environmental carrying capacity' (generally determined by biological factors, such as food supplies and/or interactions with competitors, mutualists, predators or parasites).

A K-selected organism sees its environment as relatively stable and predictable (and consequently the population is usually around its equilibrium values $N \simeq K$). This steady environment tends, however, to be biologically crowded with competitors (of the same and other species), predators and parasites. The evolutionary pressures on an organism in these circumstances are, crudely, to be a good parent and competitor, to increase the effective value of K, and to have fewer offspring but to invest more time and energy in raising them.

Conversely, an r-selected organism sees its environment as unstable and unpredictable (and is usually at low population values, growing exponentially, and undergoing episodes of boom and bust). The evolutionary pressures here are for opportunism, for large r to exploit the transient good times, and to have many offspring, few of which can expect to mature. For the r-selected organism, life is a lottery, and it makes sense simply to buy many tickets!

As emphasized above, the dichotomy of r selection versus K selection is a gross oversimplification, which deliberately polarizes what is, in fact, a complex continuum. Subject to this caveat, the ideas illuminate some of the broad trends among animal and plant species: between the r-selected insects and the K-selected mammals; between most fish (with their millions of eggs, and where next year's recruitment is roughly independent of this year's stock size) and marine mammals (where recruitment is explicitly dependent on stock size); between early successional weeds, and the trees and perennials of later successional stages. Also, as we shall see below, they help explain interesting trends and patterns within a single taxonomic class, namely mammals.

The remainder of this chapter is organized as follows. In the next section, we explore general aspects of mammalian life-history strategies. The section begins with remarks about optimal strategies; goes on to document and discuss various correlations between birth rate, generation time and physical size in mammalian species. It then looks at the way average litter size varies systematically along certain environmental gradients and concludes by emphasizing various self-reinforcing tendencies among the factors involved in r and K selection. The third section deals with aspects of the timing of reproductive events: iteroparity versus semelparity (extended versus concentrated reproductive life), litter spacing, and marsupial versus eutherian mammals. The fourth section examines patterns in the care of young: weight of newborn infants relative to parental weight, altricial versus precocial young (slow versus fast maturing), and nursing periods.

Finally, we acknowledge that Darwinian selection acts to maximize an individual's genetic input into the next generation. This means we must pay attention not just to fecundity and survival, but also to sexual selection, and to social organization and behaviour (bearing in mind that your relatives carry your genes, proportionally to how closely related they are to you). Accordingly, the fifth section discusses mating systems and social groupings.

Size and life-history strategies

Optimal life-history strategies

The notion that an organism's life-history strategy – its patterns of mortality, fecundity and parental care – depends on its physical environment and on its interactions with other organisms dates back at least to Darwin and Wallace. Recent years, however, have seen the growth of a more quantitative approach to the subject, pioneered by Edward Deevy's demonstration in 1947 that data from natural populations of non-human animals could be used to construct actuarial tables of age-specific survivorship and fecundity, similar to those constructed by insurance companies for human populations.

The most interesting of these recent studies have emphasized that the number of offspring, and their chances of survival to maturity, depend on the amount of parental investment (for example, on the number and size of eggs, and on the amount of care, if any, given to offspring), and that this in turn influences the parent's survival probability. Thus, in any attempt to calculate optimum life-history strategies for particular environmental circumstances, it must be acknowledged that age-specific mortality and fecundity schedules are not independent, but are interwoven in a complicated way; the trade-off between reproduction and 'personal growth' made by an adult in any one year will affect its probability of surviving that year, and thereby influence the possible trade-offs in all future years.

Richard Southwood discusses the problem in a way that is particularly simple and clear, yet which retains all the biological essentials. He invokes an imaginary animal, the parthenogenetic block-fish which has a productivity of two blocks each summer and one block each winter. The two summer blocks can be added to the fish itself, or used in reproduction, in any proportion (any fish that does not add to itself dies); the one winter block is necessarily added to the fish itself. What is the optimal life-history strategy for the block-fish? Obviously there is no unique answer, but rather it depends on the environmental setting. Fig. 1.2 illustrates the genealogy of two mutant strains of the block-fish: mutation 1 puts all its summer productivity into reproduction; mutation 2 puts only half its summer productivity into reproduction. If half the juvenile one-block fish are killed by predators each year, mutation 2 does better. Conversely, if the carrying capacity of the fish's environment is limited to a total of eight blocks, mutation 1 is represented at the end of 3 years by many more individuals, and is better placed to 'bounce back' from adverse environmental fluctuations. Other assumptions could clearly be explored; for example, block-fish over a certain size could exploit additional resources, or

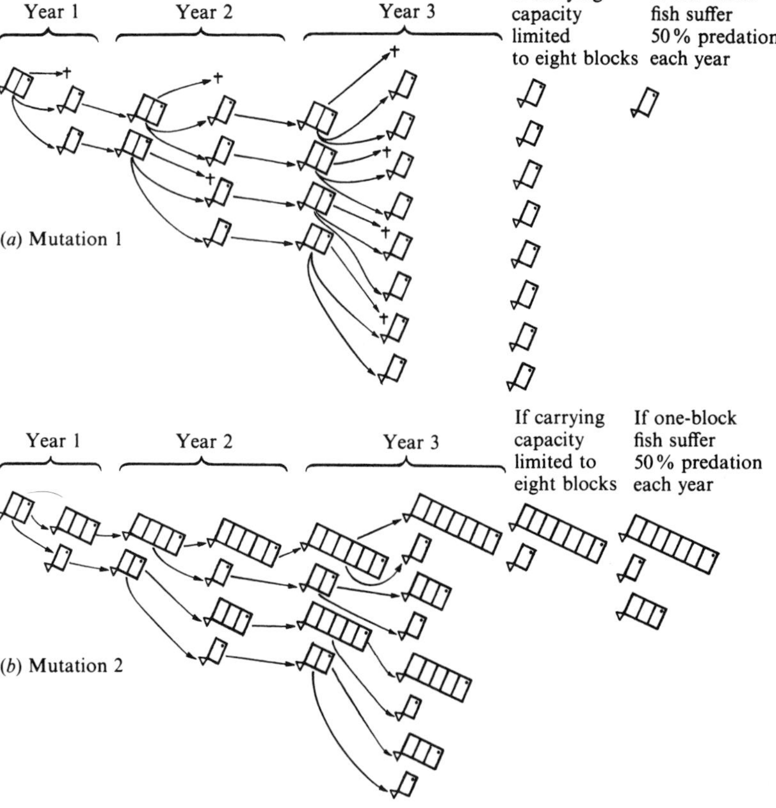

Fig. 1.2. Results of two different resource allocation strategies in a hypothetical animal, the block-fish: (*a*) mutation 1, in which summer productivity is entirely devoted to reproduction (2 blocks); (*b*) mutation 2, in which half the summer productivity is devoted to reproduction (1 block) and half is devoted to growth (1 block) and, hence, to adult survival. (From T. R. E. Southwood. Bioeconomic strategies of population parameters. In *Theoretical Ecology*, pp. 26–48. Ed. R. M. May. Blackwell Scientific Publications; Oxford (1976).)

mortality could be higher or lower for larger block-fish, or the environmental carrying capacity could be subject to random variations.

Physical size and its implications

An organism's life-history strategy and other adaptations to its environment are not, however, infinitely maleable. Rather they are confined within broad bounds by the exigencies of developmental processes and mechanical constraints associated with the design of workable living machines. It simply is not possible to evolve a creature having the size of an elephant, yet attaining sexual maturity at the age of 3 months!

Some of these 'design constraints' are well understood, others less so. Thus, the mechanical scaling laws or 'allometries' (connecting quantities such as body weight, length, brain size, weight of offspring at birth) are well documented empirically and fairly well understood theoretically. On the other hand, the correlations between dynamic variables (life expectancy, age at sexual maturity, and the like) and physical variables (body weight, length), although demonstrated in many empirical surveys, lack a definitive explanation; they are likely to be associated with developmental processes, and with the fact that smaller creatures tend to have higher weight-specific metabolic rates, living more frenetic lives and thus 'wearing out' faster.

Fig. 1.3 shows John Bonner's computation of the roughly linear relationship between generation time and body length, which extends over

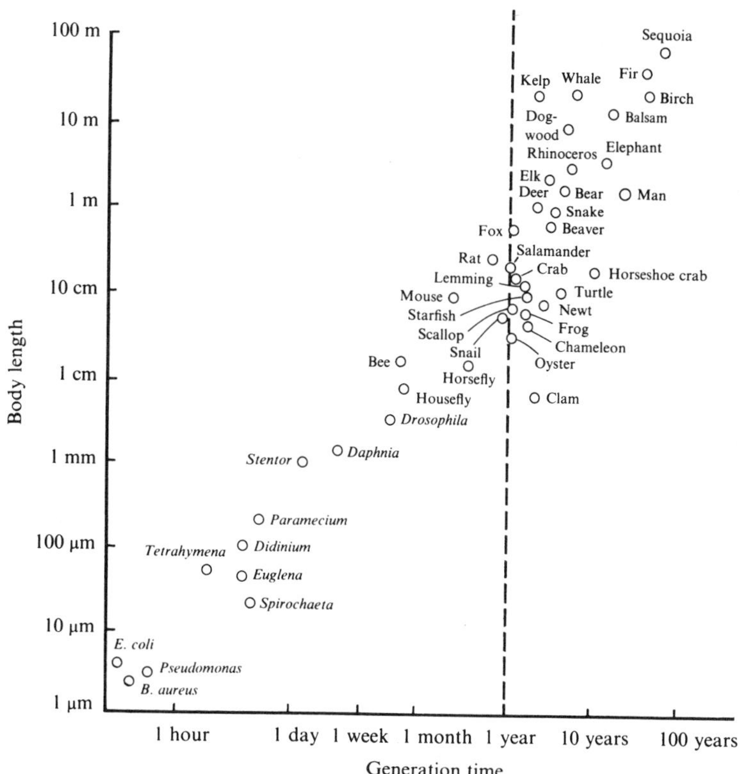

Fig. 1.3. Relationship between generation time and body length. (From J. T. Bonner. *Size and Cycle: an Essay on the Structure of Biology*. Princeton University Press (1965).)

a wide range for an extraordinarily diverse array of organisms. This general relation continues to hold when the focus is narrowed to include only mammalian species. In particular, Western has studied data for African mammals, and has shown that dynamic quantities, such as age at first reproduction, T, are empirically related to body weight, W, by scaling laws of the form

$$T = aW^b. \tag{1.1}$$

Here a and b are constants, estimated as the intercept and the slope, respectively, of the regression line when the data points are displayed on a log–log plot. African mammals can be grouped into three broad taxonomic categories: artiodactyls (cloven-hoofed), primates and carnivores. Fig. 1.4 is typical, showing the relation between age at first reproduction and body weight for African artiodactyls; in Equation 1.1 here the exponent $b = 0.27$, and similar results exist for primates and carnivores (with $b = 0.32$ in both). The analysis made by one of us (D.I.R.) pulls together information for a larger, global assembly of some 180 mammalian species, grouped into ungulates (all hoofed animals), primates, small mammals and carnivores, a sample size considerably larger than Western's. Again, log–log plots of age at first reproduction against body weight yield statistically significant relations of the form of Equation 1.1 (with slopes of $b = 0.37$ for ungulates, $b = 0.41$ for primates, $b = 0.25$ for small mammals, and $b = 0.20$ for carnivores).

The 'allometric' scaling laws that connect various physical quantities – for example, body weight W and length L – in terrestrial vertebrates can be satisfactorily explained by structural mechanics. The essential message

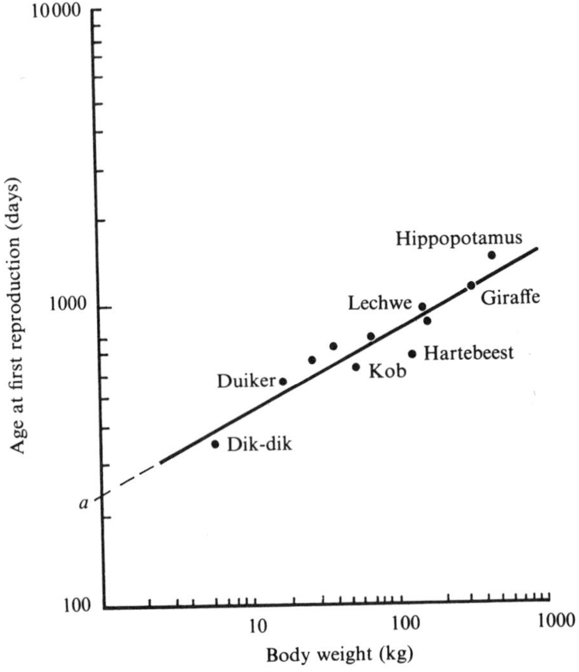

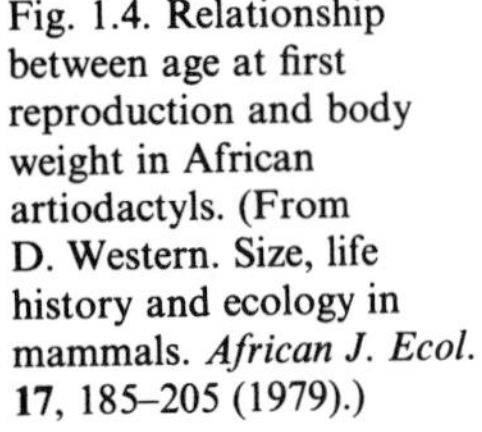
Fig. 1.4. Relationship between age at first reproduction and body weight in African artiodactyls. (From D. Western. Size, life history and ecology in mammals. *African J. Ecol.* **17**, 185–205 (1979).)

is that, for organisms large enough for gravity to be a significant factor (and this is true for all vertebrates), physical dimensions do not scale geometrically; instead, larger animals need to be relatively more squat and thick-boned to withstand the greater gravitational stresses to which they are subject. Thus, in general, body weight W tends to be related to length L as $W \sim L^4$ rather than the geometric $W \sim L^3$. Hence, Equation 1.1 corresponds roughly to $T \sim L^{4b}$ for mammals; with values of b lying in the range from 0.41 to 0.20, this gives an exponent ranging from 1.6 to 0.8 for the relation between age at first reproduction and length in mammals. This is crudely consistent with the slope $b = 1$ for the vast range of organisms in Fig. 1.3.

These scaling laws have direct implications for the population dynamics of the various species. To an excellent approximation, the intrinsic growth rate of a population, r, can be related to quantities characterizing its life-history strategy by

$$r \simeq \frac{\ln R_0}{T_c}. \tag{1.2}$$

Here R_0 is the average number of female offspring produced over the lifetime of an individual female, and T_c is the 'cohort generation time' (which is a precisely defined quantity, related to fecundity and survival schedules, but roughly corresponding to one's intuitive notion of 'generation time'). The symbol 'ln' denotes the natural logarithm. The intrinsic growth rate, r, is the effective compound interest rate at which the population is capable of growing; the population can double in $0.69/r$ years (if r is expressed as a growth rate per annum). We note that r depends only logarithmically – which is to say insensitively – on R_0. Moreover, the cohort generation time is roughly proportional to the age at first reproduction, T. Thus, to a very crude approximation, the intrinsic growth rate of a mammalian population depends simply on the age at first reproduction:

$$r \sim 1/T. \tag{1.3}$$

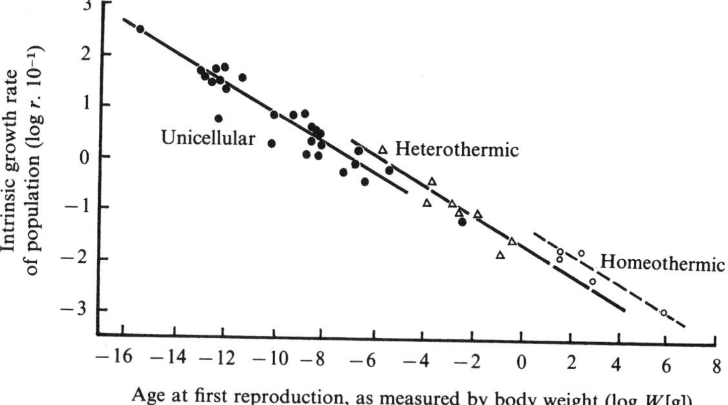

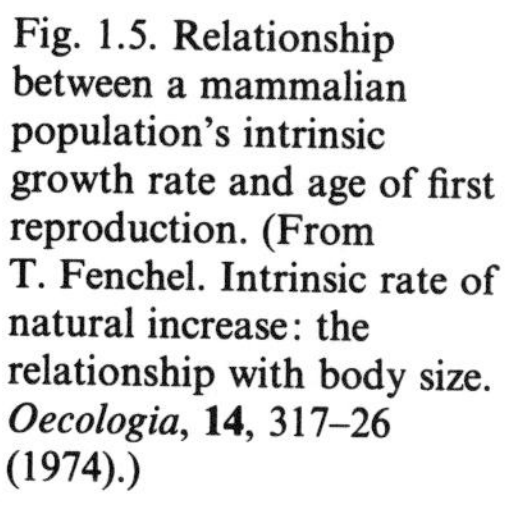
Fig. 1.5. Relationship between a mammalian population's intrinsic growth rate and age of first reproduction. (From T. Fenchel. Intrinsic rate of natural increase: the relationship with body size. *Oecologia*, **14**, 317–26 (1974).)

Combining this with the scaling law, Equation 1.1, discussed above, we have the very rough relation

$$r \sim 1/W^b. \qquad (1.4)$$

As shown by Fig. 1.5, the simple relation given by Equation 1.4 holds for an astonishingly wide range of organisms, homeothermic, heterothermic and unicellular (and with b, overall, in the vicinity of $b \simeq 0.3$). Again focusing back on mammalian species in particular, Fig. 1.6 shows the scaling law between per capita birth rates (r is the difference between per capita birth and death rates) and body weight for African mammals (here $b = 0.33$). At the lower end of the body weight series there are two closely

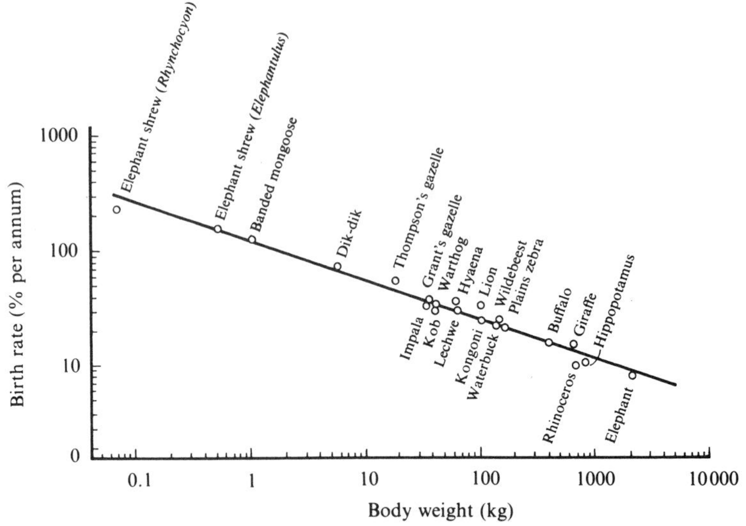

Fig. 1.6. Relationship between per capita birth rates (percentage of young born each year to the whole population) and body weight of African mammals. (From D. Western. Size, life history and ecology in mammals. *African J. Ecol.* **17**, 185–205 (1979).)

Fig. 1.7. A specimen of the rufous elephant shrew *Elephantulus*. (By courtesy of Professor John Hearn, Zoological Society of London.)

related members of a rare African rodent group, *Rhynchocyon* and *Elephantulus*. A specimen of the latter genus is illustrated in Fig. 1.7.

Litter size and environmental gradients

Within these broad scaling laws are, however, fine-grained patterns associated with local environmental or ecological factors. In other words, the design constraints discussed in the previous section set the larger patterns, which may then be fine-tuned by the specific strategic considerations discussed in the first section.

One way of seeing this is to consider, for example, how the average litter size for a particular taxonomic group varies over a latitudinal gradient. By dealing in this way with an assembly of species that have roughly similar physical sizes and behaviour, one may hope to tease apart from those differences in life-history strategies that depend on environmental differences. A classic early study of this kind was made by David Lack, who compared the clutch sizes of bird species in tropical and temperate regions. The underlying thinking is that environments tend to be more predictable and biological interactions to be relatively more important in the tropics (and make for '*K* selection', with relatively small clutch sizes and more parental care), whereas temperate environments are relatively less predictable (making for '*r* selection', with relatively large clutch sizes). The facts support these predictions. In a more detailed study, Cody took some 200 species of birds, grouped in five families, and showed that there was a significant linear regression of clutch size against latitude for each family. Cody's regression lines can be expressed in terms of their slope, s, which measures the increase in average clutch size per 1° increase in latitude; the results for birds give s around 0.06, with a range from 0.03 to 0.09, corresponding to clutch size increasing from around two to five or six, as we go from the Equator to 50° north or south latitude.

For mammals in North America, a study by Rexford Lord showed similar relations for the slope s of the regression line, expressing the increase in average litter size with increase in latitude, within each group studied. Lord's results are summarized in Table 1.1. They show significant patterns of increases in litter size along the expected '*K*- to *r*-selected' environmental gradient for six of the twelve groups; three of the remaining groups show litter size increases that are, however, not significant at the 95 per cent confidence level.

Fig. 1.8 shows average litter size as a function of body weight (plotted, as always, on a log–log scale) for a compilation of mammalian species. In this figure each species is assigned to one of three geographical zones, namely tropical, temperate or arctic, and it provides a clear demonstration of adaptative fine-structure within broad constraints (set here by body size), as discussed above. All three regression lines show an allometric tendency for average litter size to decrease with increasing adult body weight. Within this overall pattern, the three lines are clearly distinguishable;

particularly for small mammals (as for birds) there is a strong propensity towards larger litter sizes along an arctic–temperate–tropical gradient.

Life-history strategies will, of course, not depend simply on the physical environment, as crudely reflected in the latitude, but will also involve all manner of behavioural and ecological considerations. To illustrate this, we redraw Fig. 1.8, regrouping the 172 mammalian species into three new categories, according to whether they house their litters in trees (arboreal), on the ground (terrestrial or fossorial), or in burrows (Fig. 1.9). Again, all three regression lines show the general tendency for average litter size to decrease with increasing body weight, but, superimposed on this overall pattern, there is a tendency for burrowing mammals to have larger litters than ground-nesting ones, which in turn tend to have larger litters than

Table 1.1. *Litter size as a function of latitude among North American mammals. The tabulated six groups show patterns of increase in litter size with latitude that are statistically significant at the 95 per cent confidence level. Ground squirrels, pocket gophers and rats show increases that are not statistically significant, while foxes, cats and mustelids show no apparent correlation between litter size and latitude.* (R. D. Lord. Litter size and latitude in North American mammals. *Amer. Mid. Nat.* **64**, 488–99 (1960))

Taxonomic group	Number of species in study	s = increase in mean litter size per 1° increase in latitude
Rabbits	21	0.25
Tree squirrels	16	0.05
Meadow voles	18	0.13
Chipmunks	7	0.05
Deer mice	14	0.16
Shrews	12	0.12

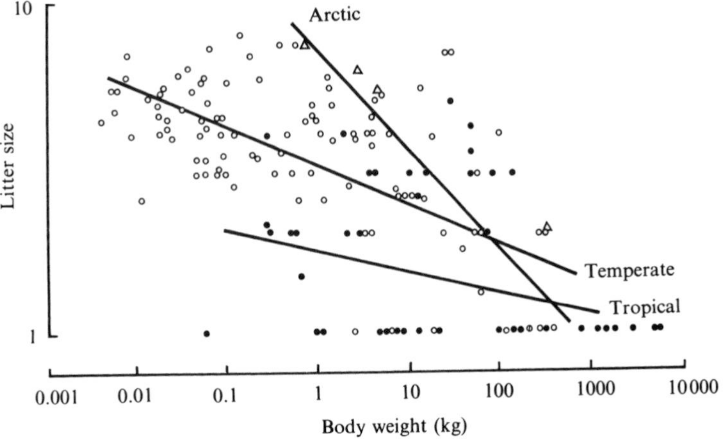

Fig. 1.8. Relationship between litter size and body weight for arctic (△), temperate (○) and tropical (●) animals. (From D. Rubenstein. *Evolutionary Ecology of Mammalian Life-histories and Social Organization* (in press).)

arboreal mammals. These statistically significant patterns may plausibly be attributed to the relatively greater and more unpredictable hazards (mainly from predation) to which burrow-dwellers are exposed, compared to tree-dwellers.

Ideally, we could attempt a three-dimensional figure, combining Figs. 1.8 and 1.9, and in which species were codified into nine classes according to geography and nesting habits. Yet other aspects of behavioural ecology could also be considered. We hope, however, that the main points are clear from Figs. 1.8 and 1.9.

Self-reinforcing feedback loops in r and K selection

Looking back on the way environmental unpredictability, life-history strategies, scaling laws and population dynamics weave together, it is clear that the notion of *r* and *K* selection has the nature of a 'Gestalt' (overall shape) rather than a simple causal chain. Thus, taking one view, a relatively small organism will have a short generation time, resulting in an intrinsically high rate of population growth, which in turn makes for a propensity to track fluctuations of the environment in episodes of outbreak and crash. On the other hand, small size and the associated short generation time condemns an organism to seeing the environment as relatively unpredictable (whereas a longer-lived organism may average-over fluctuations, or adapt to seasonal changes), which favours the evolution of a capacity for rapid population growth. In short, there are self-reinforcing tendencies, supplying positive feedback as one runs around the various causal loops. Fig. 1.10*a* illustrates these ideas about *r* selection in more detail; Fig. 1.10*b* illustrates the corresponding, but opposite, tendencies manifested under *K* selection.

We end this section with a discursive speculation. Ultimately, the set of life-history strategies for a group of species that live together will be shaped and constrained by patterns of community organization. These patterns are, at present, not well understood. Whatever the mechanisms, one outcome appears to be that there are more species of small creatures than of large ones; this is true both within broad assemblages cutting across

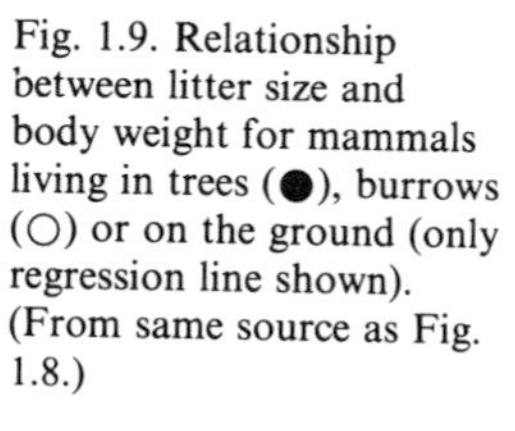
Fig. 1.9. Relationship between litter size and body weight for mammals living in trees (●), burrows (○) or on the ground (only regression line shown). (From same source as Fig. 1.8.)

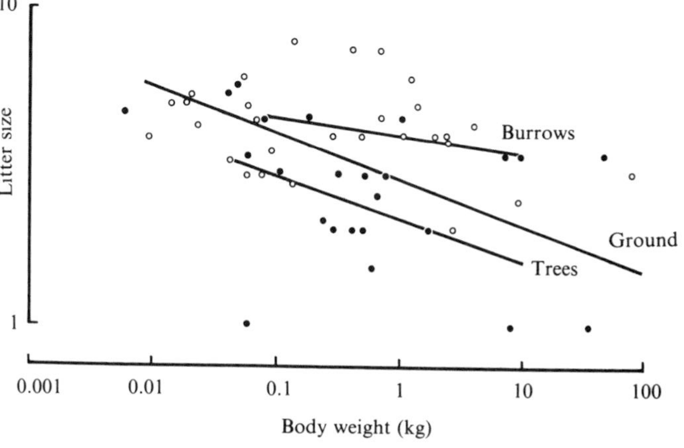

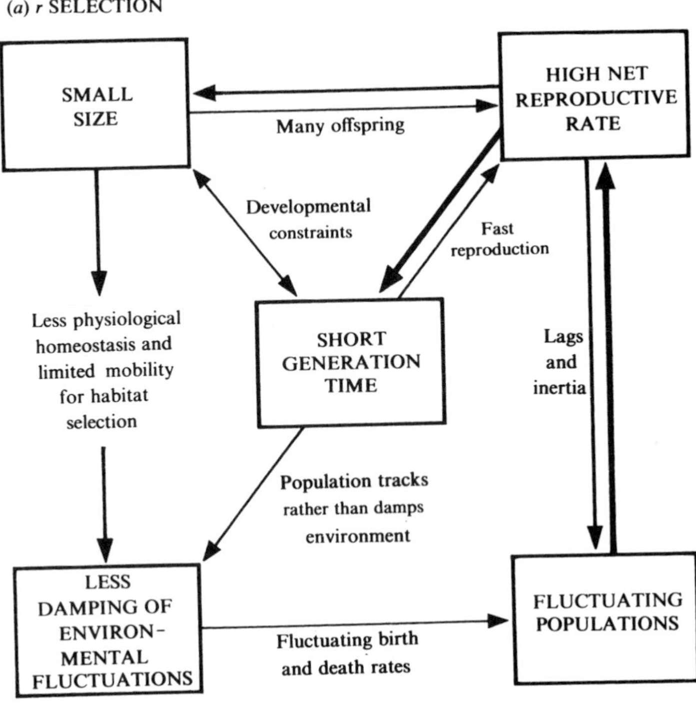

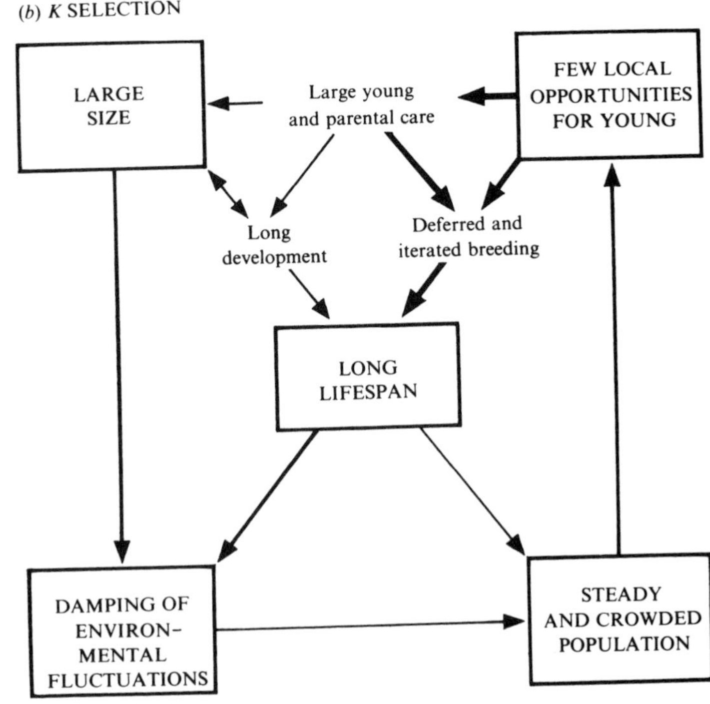

Fig. 1.10. Self-reinforcing patterns of (*a*) *r* selection and (*b*) *K* selection. Arrows point from causes to effects, and boldness depicts strength, for both. (From H. S. Horn. Optimal tactics of reproduction and life-history. In *Behavioural Ecology*, pp. 411–30. Ed. J. R. Krebs and N. B. Davies. Blackwell Scientific Publications; Oxford (1978).)

taxonomic boundaries and within particular taxa. Fig. 1.11 shows the number of species in different size categories (classified according to body weight) for the global total of mammalian species and for those mammals that currently inhabit Britain. Any eventual explanation of these patterns will, most likely, draw upon the themes developed in this chapter, but we are still a long way from such understanding.

Timing of reproduction

Iteroparity versus semelparity

One extreme and oversimplified aspect of the r-and-K-selection dichotomy centres around whether an organism's reproduction is concentrated into a single, 'big bang' event (semelparity), or distributed over a repeated sequence of breeding events (iteroparity). It can be argued that extreme forms of r selection favour rapid attainment of sexual maturity, with all available resources then channelled into one climactic production of offspring (followed usually by the death of the parent). Examples abound in many insect species, and in the celebrated self-sacrificial reproductive effort of Pacific sockeye salmon. Conversely, the K-selected paradigm is often represented as an animal maturing relatively slowly, and distributing reproductive effort over many breeding seasons once sexual maturity is reached.

Essentially all mammals are iteroparous, which suits their image as lying to the K-selected extreme among all animals. There are, however, exceptions: nine small Australian marsupials in the genera *Antechinus* and *Phascogale* are known to be semelparous. Like most exceptions, they help to define the general rule. The young of *A. stuartii* are born in September–October (early summer), weaned in December–January, and by July (mid

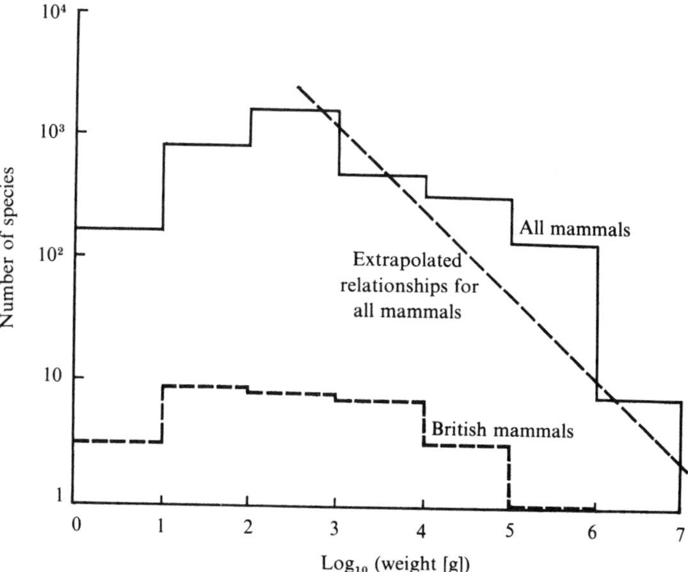

Fig. 1.11. Histogram of number of species of mammals that fall into various size (body weight) categories. (From R. M. May. The dynamics and diversity of insect faunas. In *Diversity of Insect Faunas* (Royal Entomological Society Symposium, London), pp. 188–204. Ed. L. A. Mound and N. Waloff. Blackwell Scientific Publications; Oxford (1978).)

winter) are mature. At this time males become aggressive, space themselves out and establish territories. Around August–September, females come into oestrus and are receptive for a period lasting about 2 weeks. During this period males mate repeatedly, their bodily condition deteriorates seriously, diseases set in and they all die. Babies are born 25–31 days after conception and are nursed until December–January, at which time 50–100 per cent of the adult females die.

As yet, no one has given an ineluctable reason why these species should be semelparous; they do, however, fit the criterion of being extremely r-selected, living in a forest environment where rainfall and insect abundances fluctuate seasonally and where fires are not uncommon. According to theories of life-history evolution, an individual should adopt a big-bang strategy of reproduction only if it can, by its suicidal transfer of energy, produce P/Y more young than a normal litter, where P and Y are the probabilities of the parent and young, respectively, surviving to the next year. The prolonged 5-month dependence of the *Antechinus* young on the mother leaves her nutritionally depleted, perhaps to the extent that she cannot sufficiently replenish her reserves before the onset of winter when insect abundance is low. If this were the case, P would be small, thus making it more likely that complete transfer of maternal reserves to the young would be favoured, as it would enable production of the largest possible litter and increase the survival prospects of the young. Apparently this is what occurs, as all the mother's teats are occupied and, at weaning, the weight of the litter is three times that of the mother.

The notion that strong seasonality is a force favouring semelparity is supported by the observation that sibling species of *A. stuartii* living in habitats where rainfall is seasonal or monsoonal are also semelparous, whereas those living in tropical forests are not.

Litter frequency

The iteroparous mammals exhibit a great variety of breeding patterns. To some extent, these are influenced by the physical environment. Thus, in seasonal temperate regions, and especially in boreal and arctic regions, there is a tendency for offspring to be produced in the spring; such seasonal reproduction may produce one, or more than one, litter each year. In the less seasonal tropics, breeding can generally take place at any time. These considerations suggest that there may be, on average, systematic differences between the frequency at which litters are produced in tropical versus temperate environments. The nature of any such gradient is not clear *a priori*: one line of argument suggests that the relatively 'K-selective' nature of the tropics will make for relatively lower litter frequencies; another line suggests that the less seasonal climate of the tropics permits year-round breeding, promoting increased litter frequencies.

Fig. 1.12 presents the facts, showing litter frequencies as a function of body weight for 174 mammalian species, grouped into two general classes (tropical and temperate). There is much scatter in these data points, particularly among small mammals. The regression lines for tropical and for temperate mammals both show the effects of systematic scaling, with large animals having greater spacing between litters. In addition, there is a small but statistically significant tendency for temperate mammals to have higher litter frequencies (shorter spacing between litters) than tropical mammals.

Many environmental complexities are glossed over in broad characterizations like 'tropical versus temperate'. In particular, many 'tropical' environments (for example, drought-prone regions) may be non-seasonal yet severely unpredictable. In such regions, evolutionary pressures will be geared to minimize the wastage of reproductive effort when drought unexpectedly strikes; the marsupial reproductive mode is arguably such a mechanism.

Marsupials versus eutherian mammals

Reproduction in marsupials is quite different from that of eutherians and represents a successful alternative. Marsupials are born at a very early developmental stage and make their own way to the mother's pouch while still naked, blind and very small. There they fasten to a teat and begin to suckle. In macropod (big-footed) marsupials, like the kangaroos and wallabies, a new ovum is fertilized at a postpartum oestrus. Its development is, however, arrested by the effects of the older embryo's suckling (see Fig. 2.8 in Book 2, Second Edition). In species that live in predictably seasonal environments, such as the tammar wallaby, this lactational diapause is supplemented by a photoperiodically controlled seasonal diapause. Only with the return of favourable conditions at mid summer does development of the blastocyst resume.

In more marginal desert habitats, where conditions are less predictable, a more flexible system has evolved. For both the red kangaroo and the euro, reduced rates of suckling by the joey as it starts to vacate the pouch

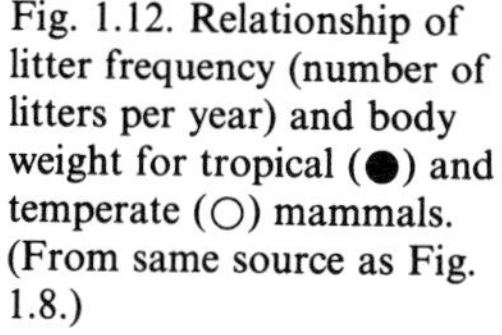
Fig. 1.12. Relationship of litter frequency (number of litters per year) and body weight for tropical (●) and temperate (○) mammals. (From same source as Fig. 1.8.)

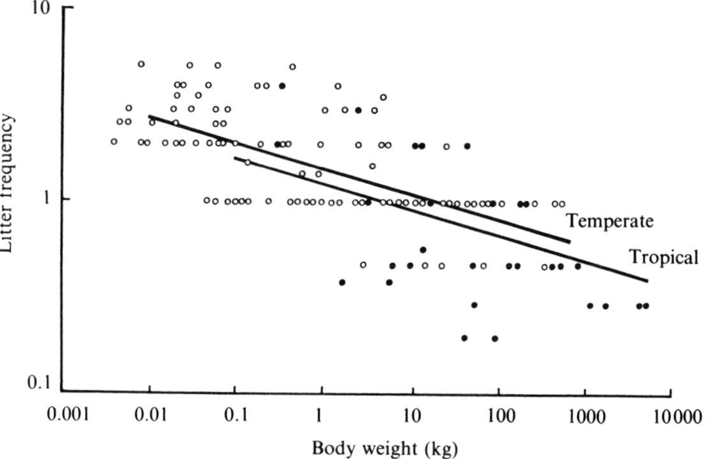

allows the blastocyst in diapause to resume development. After about a month it is born, and attaches to a teat. The mother ovulates again within a day of birth, and this new fertilized egg enters a state of suspended development in the uterus. Thus, a marsupial mother inhabiting an unpredictably changing environment may be nourishing three youngsters simultaneously, all in different stages of development: an older infant who is semi-independent but who returns to the pouch for occasional sips of milk rich in fats and proteins; a middle-aged infant permanently attached to the teat who receives a milk more deficient in these ingredients; and a real tot who is nothing more than a blastocyst, lying dormant in the uterus. This contrasts sharply with eutherian mothers who suckle only a one-age cohort of young at a time, although they may do so while a second cohort is developing in the uterus.

Such a flexible developmental system has advantages in unpredictable environments. In an area such as the Australian desert, where drought is common yet where its onset and termination are unpredictable, attempts at reproduction, if poorly timed, could be fatal to both the mother and her offspring. Rather than tying the onset of mating to environmental cues, the red kangaroo and euro mate at any time and then take advantage of their ability to arrest the initiation of development of the fertilized embryo until the pouch offspring have reduced their suckling rate. This can come about either because the offspring has matured and stands a chance of making it on its own, or because it has perished because environmental conditions have adversely affected the mother. In either case, when suckling stops, the arrested embryo in the uterus begins to develop. If harsh conditions persist, the resulting newborn may itself perish, allowing the mother to ovulate and become pregnant once more. This process could repeat itself as long as conditions remain poor, but as soon as the rains return there is always a blastocyst ready to go. Moreover, if conditions deteriorate rapidly, some (the older cohort) or all (both cohorts) of young could be jettisoned quickly and easily, with less risk to the marsupial than to a eutherian mother who would have to abort a fetus. And even after such a loss, the marsupial mother, with her waiting embryo, would be able to capitalize on any return of favourable conditions more quickly than her eutherian counterpart. By having short generations that can be prolonged at will by an older infant, the large *K*-selected marsupials inhabiting uncertain environments can become efficient opportunistic breeders.

Care of young

Birth weight relative to parental weight

One rough index of the amount of parental investment up to the time of parturition is the weight of offspring at birth relative to maternal weight (often called the 'proportionate birth weight'). Fig. 1.13 shows this proportionate birth weight as a function of maternal weight for some 69 species of hoofed mammals. Ungulates are especially interesting because they inhabit a wide range of environments and comprise species in many different stages of domestication.

Fig. 1.13 shows that the maximum relative reproductive effort among ungulates apparently occurs in pronghorn antelope and the goat, at around 15 per cent, decreasing to a minimum of around 4–5 per cent for the very large ungulates and subungulates. The trend to diminishing proportionate birth weight with increasing maternal weight is fairly steady for the larger animals, but there is much variability among the small species (adult weights less than or around 100 kg). Although there is too much scatter for any crisp generalization, Fig. 1.13 shows that several north temperate and arctic ungulates (for example, caribou, muskox, bighorn sheep) tend to produce relatively smaller offspring than most African ungulates, with other North American species typically being intermediate.

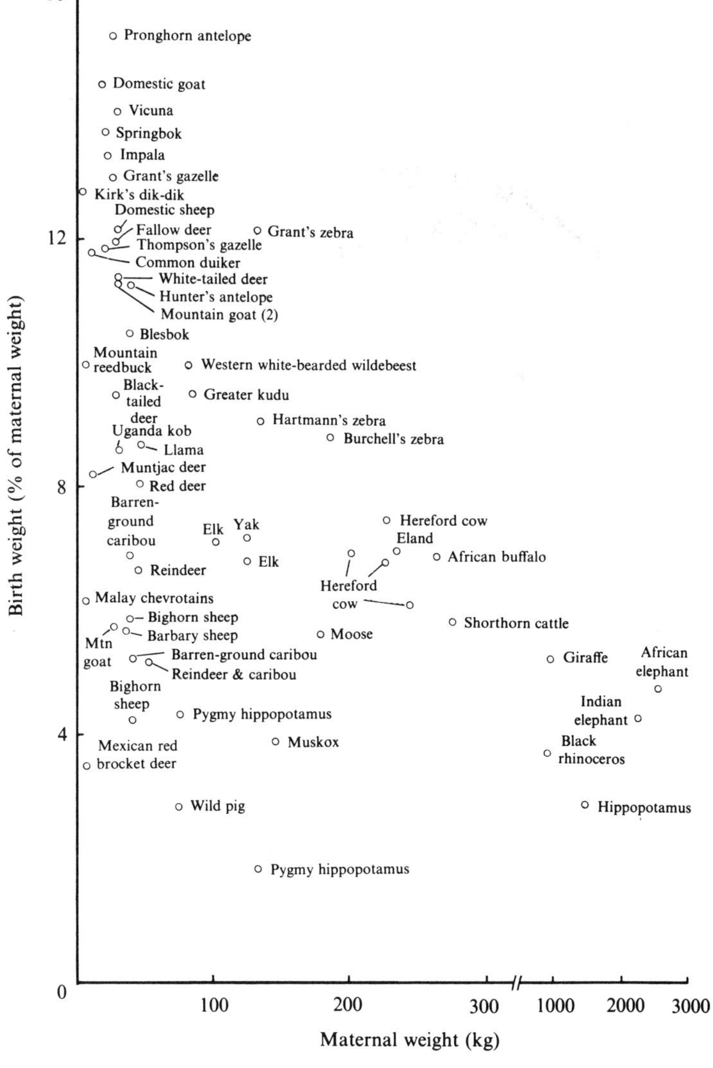

Fig. 1.13. Relationship between birth weight and maternal weight among hoofed mammals. (From C. T. Robbins and B. L. Robbins. Fetal and neonatal growth patterns and maternal reproductive effort in ungulates and subungulates. *Amer. Nat.* **114**, 101–16 (1979).)

The explanation for the relatively small single fetus of caribou, muskox and bighorn sheep may lie in inadequate food supplies during winter, and the need to optimize maternal survival relative to reproductive effort; these factors may override thermodynamic considerations, which would argue for the largest possible neonate, to minimize heat loss in a cold environment.

Revealingly, there are yet further complications among the smaller species represented in Fig. 1.13. The Mexican red brocket deer, a shy tropical animal which has one of the lowest proportionate birth weights of the smaller ruminants, has a postpartum oestrus. So do muntjac deer and several African bovids. These species live in relatively stable environments, and the combination of relatively low birth weight and postpartum oestrus enables them to distribute reproduction throughout the year, in a way that may well increase total reproductive output.

In short, Fig. 1.13 again shows that there tend to be systematic scaling patterns, which are common to natural and domesticated species. Within these broad tendencies, particularly for the smaller species, there resides a fascinating richness of detail that overwhelms any easy generalization.

Altricial versus precocial young

The relative size of offspring at birth is, of course, only part of the story of parental care. Some young are born relatively independent; such 'precocial' young are relatively able to thermoregulate, and to forage and fend for themselves. In contrast, 'altricial' young require significantly more parental care in the first stages of their life. Since most of the energy demands of altricial young must be met by the parents, such offspring cost their parents more in terms of total expenditure of energy than do precocial young. Thus, only under special conditions will it be 'cost effective' for a species to have a relatively short pregnancy resulting in the production of altricial young.

One such condition arises when the postpartum rearing environment

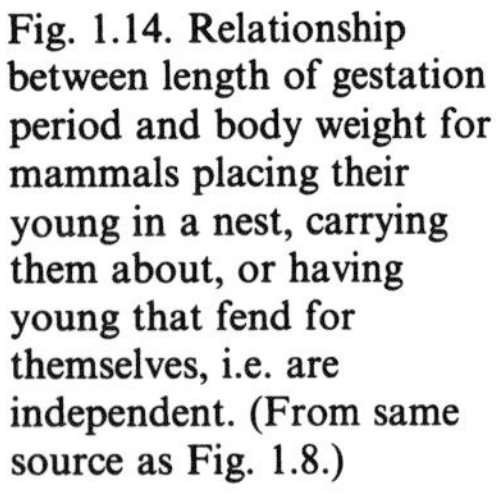

Fig. 1.14. Relationship between length of gestation period and body weight for mammals placing their young in a nest, carrying them about, or having young that fend for themselves, i.e. are independent. (From same source as Fig. 1.8.)

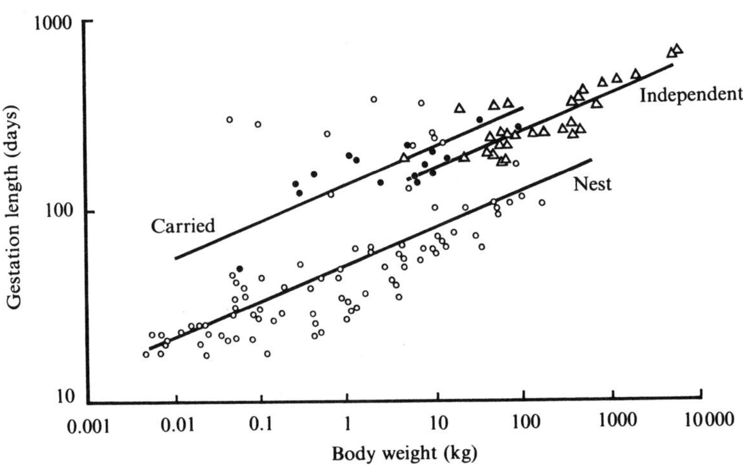

offers more security than the womb. All else being equal, selection will favour depositing the young in such safe places as early as possible. As Fig. 1.14 shows, mammals placing their young in nests have shorter gestation periods than those carrying their infants about, or those whose young are independent from birth. A second condition favouring the production of altricial young arises when juvenile adjustments to the vagaries of feeding and competing with adults, or socializing, require so much skill and physiological maturity that only an extraordinarily and unobtainably long period of gestation would be required. Since catching large prey is not an easy task for youngsters or pregnant females – in contrast to grazing on grass or browsing on herbs (which are more abundant and continuously distributed) – we would expect carnivores to bear relatively smaller young than ungulates. Fig. 1.15 shows that this is indeed the case.

Whether or not a female should bear relatively large young should also depend on whether or not she can offset the investment of a disproportionate

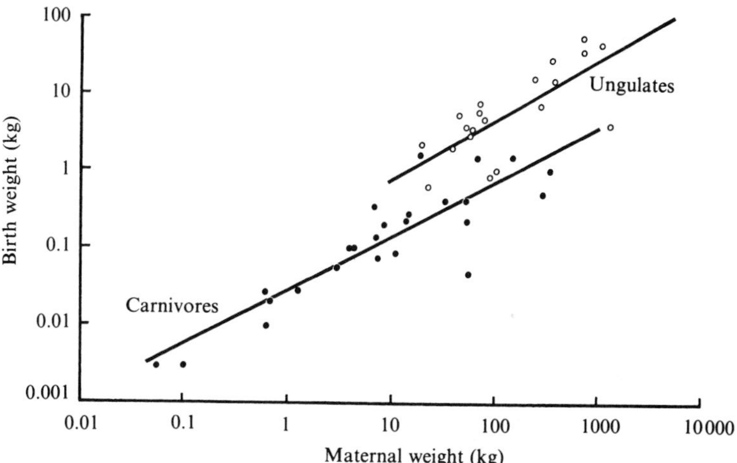

Fig. 1.15. Relationship between birth weight and maternal weight for ungulates and carnivores. (From same source as Fig. 1.8.)

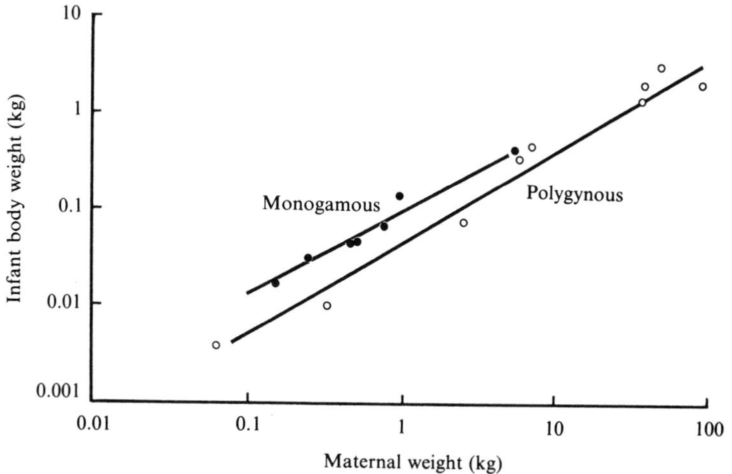

Fig. 1.16. Relationship between infant birth weight and maternal weight for monogamous and polygynous primates. (From same source as Fig. 1.8.)

share of her reserves in current reproduction by sharing the demands of lactation and postpartum care with others not experiencing such large physiological stresses. Since males of monogamous species assist in carrying the young or provisioning the young or its mother, we would expect monogamous mammals to produce larger young than polygynous ones. This is often the case, and is clearly portrayed by the primates in Fig. 1.16.

Mating systems and social groupings

Strategies for bearing and caring for young are certainly essential features of reproduction, but they represent only half the story. Activities associated with mating are equally important; again, the patterns adopted by particular species are influenced by features of the environment.

The asymmetries inherent in sexual systems are such that one sex usually ends up investing more of its resources than the other sex in the rearing of young. In mammals this disproportionate parental investment is usually performed by females. Since females cannot rear offspring as quickly as males can father them, selection will favour female investment strategies that maximize the number of young raised to maturity. As a consequence, female behaviour will be dominated by activities associated with acquiring food or avoiding predators. For males, however, selection will favour strategies that maximize the number of females mated; searching or competitive activities such as guarding females or defending a territory are likely to predominate in their behavioural repertoire. Apparently, the invention of lactation did little for the home life of mammals; most mammals exhibit polygynous breeding systems.

Polygyny can take a variety of forms because the success of males in gaining mating access to more than one female ultimately depends on how the environment affects patterns of female social groupings. If critical food resources are distributed unevenly in patches and are of low quality, then, on average, patches will not reliably be able to support female groups with permanent membership. Without groups to defend, males can become polygynous only if they can exclude other males from an area comprising one or more of the patches that the females require. Grevy's zebras and langurs both seem to practice this form of 'resource defence polygyny'.

When resources are distributed in patches of higher quality, large aggregations of females can regularly be supported. Most likely these will meld into coalitions with fixed membership, as intergroup competition will be intense. Once such groups are formed, superior males will be able to defend them against other males and thus gain exclusive mating access ('female defence polygyny'). Gelada and hamadryas baboons, Burchell's zebra, impala and scores of other mammals exhibit such harem breeding systems.

Sometimes males will not be able to exclude all other males from these female associations; this might happen when female groups are so large that pressure from excluded males is tremendous, or when females encourage males to immigrate as allies in their intergroup struggle. In the presence of other males, superior males will attempt to maintain their priority of access to females by aggressive domination of subordinates ('male dominance polygyny'). Yellow and olive baboons, macaques and many middle-sized ungulates exhibit these multi-male, multi-female groups. Here the reproductive success of dominant males will depend somewhat on the extent to which female reproductive cycles are synchronized and on the length of courtship.

When resources are evenly distributed, competition for any individual resource item is lessened. As a result, tight-knit female associations may not be selected, for when this occurs males will have to wander about searching out solitary females or temporary aggregations of females. A female in oestrus is likely to be found and contended for by many males (unless the pool of males is small and the dominance relationships among them are well established); here extreme polygyny may be possible to achieve and may give way to promiscuity. Large grazing mammals, such as African elephants and Cape buffalo, that tend to perceive grasslands as a more-or-less continuous sward, exhibit this type of wandering polygyny. Sometimes, however, females inhabiting these even stands of densely packed resources form groups because they help to protect against predators. At other times, food resources are not the critical resource but rather sites for resting or breeding are limited or patchily distributed, which makes it necessary for females to aggregate. In such instances, males should attempt to form harems.

Not all evenly distributed resources need consist of densely arranged prey items, and when resource items are sparse females may have to range over large distances to obtain them. In this case, groups will not form, and males will have to search widely for solitary females. Perhaps this explains why some insectivores, carnivores and marine mammals are often widely dispersed and breed promiscuously.

Breeding and social systems are not strictly governed by the distribution of critical resources such as food or nest sites. Sometimes the potential for polygyny is present and not realized, as is often the case where the survival prospects of the young can be greatly augmented by the assistance of two parents, and where female receptivity is highly synchronized so that philandering by males is precluded. Under these conditions males that abandon polygyny and mate monogamously will be favoured. In most monogamous species males share some of the parental responsibilities by defending the young, bringing food to the female or the young, or carrying the young. The adoption of parental duties by males is clearly evidenced in the monogamous arboreal primates (such as the siamang, marmosets and tamarins), where the weight of the young is large relative to that of the mother, and where the young are carried for extended periods; the contrary is the case among polygynous primate species. And, as discussed earlier, the monogamous carnivores (such as canids), which bear large litters consisting of many altricial young, assist in meeting the large energetic demands of a litter by augmenting both the mother's and the offsprings' diet by regurgitating captured prey.

In summary, resource quality and dispersion will in general affect a female's reproductive success, and thus play a central role in determining adaptive patterns of association among females. These in turn determine the mating strategies of males. When the two are taken together, the basic mammalian breeding systems and social systems emerge.

Obviously deviations from these ideal patterns do occur. Some can be accounted for by examining other important ecological variables that have been omitted in the above analysis. Others, however, can be accounted for only by expanding the analysis to incorporate the proposition that not all individuals in a population should adopt the same strategy. Since the expected reproductive success associated with one individual's mating activities depends to a large extent on what other individuals in the population are doing, the rewards of any particular strategy will often be frequency-dependent. As a result it is highly likely that equally successful alternatives (polymorphisms) will occur. Thus it is not surprising to find that in some populations of horses, gelada baboons and red deer some males establish harems and guard females, while others will be sneaky raiders attempting to steal copulations surreptitiously. Likewise, in some populations of elephant seals and of impala, some males establish territories while others become satellites, mating with the females before they arrive at the territory. Clearly, much more work is needed to identify the reasons for the striking diversity of mammalian mating patterns. The foregoing discussion, however, shows that environmental quality and predictability have an important influence on patterns of breeding and sociality among mammals.

We see that no single, crisp pattern emerges from our survey of the kaleidoscope of factors that work together to fashion the reproductive strategies of mammals.

One broad pattern that is, however, discernible is the existence of design constraints on the construction of living machines, which makes for systematic correlations between the physical size of a mammal and the magnitude of various quantities that characterize its reproductive behaviour (such as litter size and weight, or time to reproductive maturity). Within any one size class there are finer patterns, moulded by the ecology and environment for the individual species, and having to do with its reproductive biology, its mating system and its social organization. Ultimately, the physical size of the animal of a given mammalian species is itself the product of natural selection, subject to the restrictions imposed by the environment and by interactions with other species in the community.

Suggested further reading

A general theory of clutch size. M. L. Cody. *Evolution*, **20**, 174–84 (1966).

Ecology, sexual selection and the evolution of mating systems. S. T. Emlen and L. W. Oring. *Science*, **197**, 215–23 (1977).

The social organization of antelope in relation to their ecology. P. J. Jarman. *Behaviour*, **48**, 215–67 (1974).

Environmental uncertainty and the parental strategies of marsupials and placentals. B. S. Low. *American Naturalist*, **112**, 197–213 (1977).

Parental investment: a prospective analysis. J. Maynard Smith. *Animal Behaviour*, **25**, 1–9 (1977).

Ecological strategies and population parameters. T. R. E. Southwood, R. M. May, M. P. Hassell and G. R. Conway. *American Naturalist*, **108**, 791–804 (1974).

An ecological model of female-bonded primate groups. R. W. Wrangham. *Behaviour*, **75**, 262–300 (1980).

On the evolution and adaptive significance of postnatal growth rates in terrestrial vertebrates. T. J. Case. *Quarterly Review of Biology*, **53**, 243–81 (1978).

Evolutionary rules and primate society. T. H. Clutton-Brock and P. H. Harvey. In *Growing Points in Ethology*, pp. 195–237. Ed. P. P. G. Bateson and R. A. Hinde. Cambridge University Press (1976).

Life tables for natural populations of animals. E. S. Deevy. *Quarterly Review of Biology*, **22**, 283–314 (1947).

Scaling physiological time. T. A. McMahon. In *Some Mathematical Questions in Biology*, pp. 131–61. Ed. G. F. Oster. American Mathematical Society; Providence, R.I. (1980).

On the evolution of alternative mating strategies. D. I. Rubenstein. In *Limits to Action: the Allocation of Individual Behavior*, pp. 65–100. Ed. J. E. R. Straddon. Academic Press; New York (1980).

Life-history tactics: a review of the ideas. S. C. Stearns. *Quarterly Review of Biology*, **51**, 3–47 (1976).

The Mammalian Radiations. J. F. Eisenberg. University of Chicago Press (1981).

Ecological Adaptations for Breeding in Birds. D. Lack. Methuen; London (1968).

The Theory of Island Biogeography. R. H. MacArthur and E. O. Wilson. Princeton University Press (1967).

Life of Marsupials. C. H. Tyndale-Biscoe. Edward Arnold; London (1973).

The Natural History and Antiquities of Selbourne. G. White. B. White & Son; London (1789).

2
Species differences in reproductive mechanisms

R. V. SHORT

Readers of the books in this series, and particularly of this volume, will have been struck by the enormous diversity of reproductive mechanisms across mammalian species. About the only common factor is that, somewhere along the line, a spermatozoon meets an egg and so a new individual is formed. Everything else is subject to enormous variability, so that generalizations become impossible. This may act as a deterrent to the faint-hearted, who are baffled by complexity; in reality, it should be seen as an exciting challenge, encouraging us to probe deeper for the reasons underlying the diversity. Why has Nature chosen to play so many enigma variations on the theme of reproduction?

Choosing how, when and where to reproduce is the ultimate form of gamesmanship. Of one thing we can be absolutely certain: all species will have taken the utmost care in arriving at that decision. It represents the key strategy for survival, and like all strategies it will be a compromise between risks and benefits. We have now done enough detective work to provide adequate descriptive accounts of the principal reproductive events in the life of many mammals, i.e. gestation length, litter size, birth weight, age at puberty, mating system, length of oestrous cycle, breeding season, birth interval and reproductive lifespan. The challenge that lies before us is to gain a deeper insight into the environment in which the species evolved, so that we can begin to understand how the decisions were made about maximizing each of these variables.

Let us take the example of our own species. Why do we have the longest gestation of any primate, even though we are not the largest? Why are our babies so immature at birth, and why do we have the latest age at puberty of any mammal? Our anatomy would suggest that we are polygynous (males 20 per cent bigger than females), and yet many cultures advocate monogamy; is serial monogamy a comprise between these two extremes? Why do women have menstrual as opposed to oestrous cycles? Has the abandonment of oestrus been a key event in our social evolution? Does the fact that we are not seasonal breeders reveal a tropical ancestry? What was the optimal birth interval for our hunter–gatherer ancestors, and how was this achieved? And finally, why should the menopause occur, and be confined to the female? We cannot begin to answer any of these questions by studying present-day Western societies, and yet answer them we must if we are to understand ourselves as a species. These questions challenge the intellect. No clues to the answers can be found in textbooks of obstetrics and gynaecology; instead, we need a good anthropological account of human hunter–gatherer lifestyles.

When dealing with species differences, the temptation in the past has always been to adopt a *Guinness Book of Records* approach. Isn't it fascinating that the smallest mammal at birth is the honey possum, which weighs about 4 mg, and the largest is a blue whale, which weighs about 2500000 g? Did you know that in the plains viscacha up to 800 eggs are ovulated at one oestrus, that in the tenrecs fertilization occurs within the Graafian follicle, that in horses unfertilized eggs remain trapped in the Fallopian tube, that the European hare ovulates and can conceive during pregnancy, and that the giraffe fetus ovulates *in utero* and is born with corpora lutea in its ovaries? The problem with this type of approach is that we cannot begin to understand the reasons for these bizarre reproductive mechanisms, presumably because we do not know enough about the ecological constraints under which they evolved.

In order to discuss species differences in a meaningful way, we therefore need to develop a broad ecological frame of reference, and begin with some generalizations.

As Bob May and Dan Rubenstein have pointed out in the preceding chapter, the concept of *r* and *K* selection, depicting two basically different types of reproductive strategy is at best a crude generalization. In practice

Fig. 2.1. *r* and *K* selection.

MAMMALIAN REPRODUCTIVE STRATEGIES

r STRATEGISTS	*K* STRATEGISTS
Maximize their rates of population increase	Keep their rates of population increase in balance with the carrying capacity of the habitat
Colonize unpredictable environments 'Boom or bust'	Colonize stable environments and compete for resources
Typically they have: Small body size Short lifespans Short gestations Large litters Rapid rates of development Short birth intervals	Typically they have: Large body size Long lifespans Long gestations Single offspring Slow rates of development Long birth intervals

there is a no clear-cut distinction between the *r*-strategists, who tend to colonize unstable environments and do all in their power to maximize their rates of population increase, and the *K*-strategists, who try to keep their rate of reproduction in equilibrium with the carrying capacity of the habitat; what we see in fact is a continuum. Much of the apparent difference between the two extremes may in any case be little more than a consequence of differences in body size; in large animals, all reproductive events are inevitably slowed down. Nevertheless, the concept of *r* and *K* selection makes a useful starting-point for discussing the different physiological mechanisms that may be at work to regulate rates of reproduction in different species (Fig. 2.1).

Ovulation rate, embryonic mortality and litter size

For the marine biologist, accustomed to a situation in which female fish lay tens of thousands of eggs at spawning time in order to ensure that two or three of the fry will survive to reproductive age and so perpetuate the species, mammals must seem remarkably efficient in their reproduction. Although the mammalian ovary, like the fish's ovary, also contains millions of oocytes (at least initially), it has been necessary for mammals to develop mechanisms for drastically reducing the number of eggs shed at ovulation, otherwise the overcrowding that would occur in the narrow confines of the uterus could hazard normal embryonic development. We can still find a few fishy exceptions amongst mammals, the plains viscacha being a good example. But since mammalian eggs are so small, they are energetically very cheap to produce, and it matters little if many are wasted. Perhaps we should regard ovarian atresia, and the endocrine mechanisms responsible for selection of a limited number of follicles for ovulation, as secondary adaptations forced upon mammals as a result of their terrestrial environment.

The remarkable thing about most small mammals is that ovulation, egg pick-up, fertilization and implantation are all such highly efficient processes; there is usually relatively little discrepancy between counts of corpora lutea, reflecting the number of eggs ovulated, and counts of embryos; in rats and mice, about two-thirds of the eggs shed at ovulation result in viable embryos. Embryonic mortality can be increased by uterine overcrowding, such as occurs following experimental superovulation or the experimental transfer of an excessive member of fertilized eggs to the uterus. This makes the point that uterine size is an important factor for embryo survival, and it must have evolved in parallel with the ovulation rate. The uterus simplex of primates, which is all body and no horns, can accommodate only a very limited number of embryos, and then only because of the compactness of the primate placenta. There is no way in which a primate's uterus could encompass the much larger placenta of an ungulate. The extreme situation is seen in the pig, whose enormously long uterine horns are necessary to provide sufficient surface area of endometrium for the attachment of a dozen embryos, each enveloped in a voluminous epithelio-chorial placenta (see Fig. 2.2). This has even had repercussions for the boar, since he must dilute his spermatozoa with 0.5 l or more of accessory fluids, produced by his large seminal vesicles and bulbo-urethral glands, in order to ensure that some of the ejaculate is transported by peristalsis up to the tip of each uterine horn. This is a classical example of reproductive strategy determining reproductive anatomy. Wild pigs must have opted for *r* selection in the first place for purely ecological reasons, and their high prolificacy would therefore have made them particularly attractive to early man for domestication. He has since improved on Nature still further by selecting domestic pigs for larger body size, higher ovulation rates, increased litter sizes, more nipples and faster growth rates, so that pigs are now amongst the most *r*-selected of all large mammals. The fact that embryonic mortality is about 40 per cent might suggest that selection has reached the point where the ovaries are now able to shed more eggs than the uterus can accommodate.

The classical view of embryonic mortality is that it represents an unnecessary wastage of normal embryos; however, environmentally induced variability in the extent of embryonic mortality may be one way in which some species can regulate their fertility. In the rabbit, for example,

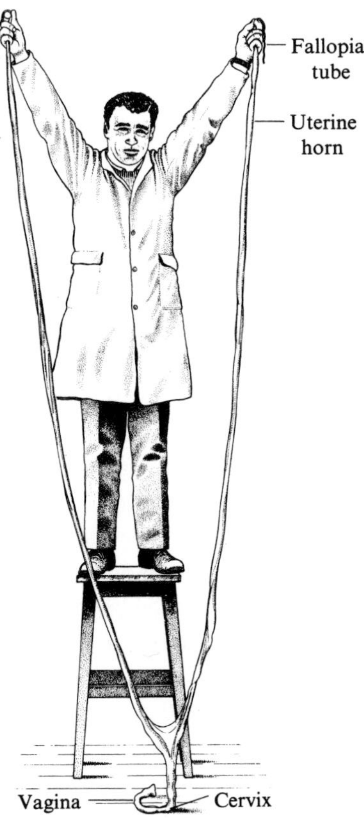

Fig. 2.2. The uterus of a pregnant sow after the embryos have been removed, showing the enormous length of the two uterine horns. The cervix is touching the ground.

if a doe is allowed to become pregnant immediately after parturition when she is suckling a large litter of more than three or four young, all the new embryos will die at the blastocyst stage. In Rogers Brambell's classical studies of embryonic mortality in wild rabbits in Wales during the Second World War, he found that mortality was greatest in does of lower body weight, and in does that were actively lactating; at least 11 per cent of eggs shed were lost in this way. Similar embryonic mortalities have been recorded in feral rabbits in Australia and New Zealand.

In addition to this environmentally induced loss of normal embryos, some embryonic mortality is caused by defects in the embryos themselves, and strange though it may seem, this could also confer certain advantages. Readers of Books 1 and 2 (Second Edition) will recall that mammalian gametes show no evidence of haploid gene expression, and yet we know that, in man at least, meiotic errors occur quite frequently during gametogenesis, resulting in the production of genetically unbalanced spermatozoa and ova. Any such defect must remain concealed within the gamete, because its genotype is not expressed in its phenotype. It is not until after the embryo has started to cleave that the maternal and paternal genomes become fully expressed; the mother's surveillance system, which enables her to detect and then reject abnormal embryos, probably does not become fully operative until after the time of implantation. Therefore, we should also look upon embryonic mortality as an effective means of filtering out abnormal embryos. Embryonic mortality might serve a particularly useful purpose in highly *K*-selected species, where prolonged birth intervals of several years are necessary for normal infant development. Any mechanism that helps to increase the interval between successive births might, within limits, prove to be highly advantageous. Thus a sprinkling of genetically defective gametes that produced a measure of non-recurrent infertility through early embryonic death might be a cost-effective way of buying time and postponing the next birth by a month or so. If the proportion of defective gametes should become too high, resulting in excessively long birth intervals, this would be subject to heavy negative selection pressure, since it would begin to pose a threat to survival.

We do have evidence that in some highly *K*-selected species, such as ourselves, the probability of conception at a given ovulation is very low. Demographers use the term 'fecundability' to express the percentage of women who conceive per menstrual cycle during which intercourse occurs, and the maximum recorded fecundability is only 28 per cent (see Table 2.1). Fecundability is strongly influenced by intercourse frequency, and also by maternal age; it is highest in women in their 20s having intercourse many times a week. It is possible to get good fecundability estimates from historical data by studying the time elapsed between marriage and birth of the first child, particularly in Catholic communities where premarital intercourse was less likely to have occurred, and where fertility-limiting practices such as coitus interruptus, abortion and infanticide would probably not have been used.

From all this information, it seems highly probable that three out of four ovulations in women having frequent intercourse fail to result in a normal pregnancy; the most likely explanation is that this is because of a very high rate of embryonic mortality, particularly during the early stages of gestation. A recent study of women who were trying to become pregnant supports this view; 43 per cent of women who were 'biochemically pregnant', with detectable levels of human chorionic gonadotrophin on day 21 of the menstrual cycle, subsequently menstruated at the normal time and thus failed to show any clinical evidence of a pregnancy.

Readers of Pat Jacob's chapter in Book 2, Second Edition, will have been impressed by the fact that 50 per cent of spontaneous abortions of clinically recognizable human pregnancies are due to major chromosomal abnormalities. The commonest of these is trisomy, when one or other of the gametes contributes an extra autosome. For each gamete with an extra autosome, there is likely to be one missing an autosome, which would give rise to a monosomic embryo. Thus, it is probable that there are as many autosomal monosomies as there are trisomies, but the fact that they have never been identified in spontaneous abortions suggests that they must be causing embryonic mortality at a very early stage of gestation, before the pregnancy becomes clinically recognizable.

This low inherent fecundability of our own species has considerable practical implications for those involved in the treatment of infertility. For

Table 2.1. *Fecundability estimates for different human populations, expressed as the percentage of women becoming pregnant per menstrual cycle during which intercourse occurs*

Population	Fecundability (%)
Historical populations	
Fifteen English parishes	22
Thirteen German villages	23
Two Belgian communities	28
Three French communities	23
Present day	
Four Latin American countries	18
Taiwan	22
North American Hutterites, aged 21	28
France, aged 25	27
United Kingdom	21
United States of America	25

(Data from R. V. Short, *Ciba Foundation Symp. Maternal Recognition of Pregnancy*, **64**, 337–94 (1979); J. Bongaarts, *Center for Policy Study*, **89**, 1–44. The Population Council; New York (1982).)

example, the success rate of *in vitro* fertilization and embryo transfer techniques is unlikely to exceed about 25 per cent, although one way of increasing it is to transfer more than one embryo at a time. Clinicians in the past have devoted a great deal of therapeutic ingenuity to the treatment of threatened and habitual abortion, fortunately to no avail. In the light of what we now know, such treatments can be regarded not only as futile, but also as contraindicated. The non-recurrent nature of much of this infertility should also give hope to the infertile couple; just because they have failed to achieve a pregnancy after a year does not necessarily mean that there is anything wrong with them.

Another highly *K*-selected species that stands out as being relatively infertile is the elephant. The ovaries of pregnant elephants invariably contain numerous corpora lutea of varying sizes, but there is usually only a single fetus. At first sight this would suggest that the elephant is in the

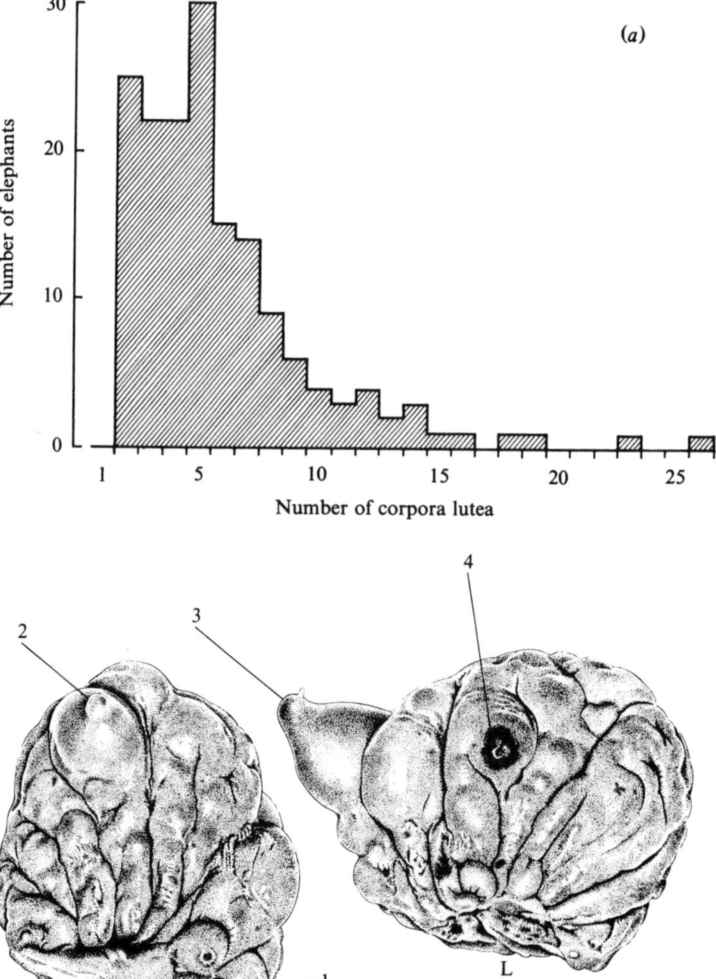

Fig. 2.3. (*a*) Histogram to show the number of corpora lutea found in the ovaries of pregnant African elephants. (*b*) The ovaries of a non-pregnant African elephant at the end of oestrus, showing a single ovulation (4) and three corpora lutea of varying sizes from previous cycles.

same league as the plains viscacha, and polyovular. However, I was able to follow a wild African elephant in Uganda throughout its period of oestrus, when it copulated with a number of different bulls. After all mating activity had ceased, the animal was shot, and I was amazed to discover in the ovaries a number of old corpora lutea, but only *one* fresh ovulation point (see Fig. 2.3). This suggests that elephants are monovular, but polyoestrous; unlike other mammals, they keep accumulating corpora lutea in their ovaries until they eventually become pregnant. Since the length of the oestrous cycle in the Asiatic elephant is now known to be about 18 weeks, each oestrus and ovulation that fails to result in a conception buys the animal an additional 4 months of infertility to contribute to the normal 4-year birth interval. Whether this low fecundability of the elephant is ultimately due to fertilization failure or early embryonic mortality remains to be determined.

Although we have cited only a few highly selected examples to prove the point, it does seem as if the small, *r*-selected mammals are characterized by high ovulation rates, relatively low embryonic mortality and large litter sizes, whereas the large, *K*-selected mammals have low ovulation rates, high embryonic mortality, and usually only a single offspring.

Gestation length

One obvious way of increasing or decreasing an animal's reproductive potential is by altering its gestation length, and the variability between species is enormous. Consider the extremes: the shortest known gestations of 12.0–12.5 days are found in the bandicoots, rabbit-sized marsupials from Australia, and the longest, about 22 months, are found in the Asiatic and African elephant. Although all the marsupials give birth to extremely small, altricial young (see Fig. 2.4), having traded gestation for lactation, it would be wrong to conclude that this is why they have short gestations. Some of the smaller marsupials have gestation lengths that are actually longer than their eutherian counterparts. For example, the marsupial shrew *Antechinus stuartii*, weighing about 28 g, gives birth to a litter of 16-mg offspring after a gestation of about 30 days, in contrast to the 1200-mg offspring of *Mus musculus*, the laboratory mouse, produced after a gestation of only 19 days (Fig. 2.5). However, the larger marsupials certainly do have very short gestations relative to their body size.

Amongst the eutherian mammals, gestation length is related to maternal body size, but there are three groups that stand out: the whales have relatively short gestations, and the hystricomorph rodents (guinea-pigs, coypus, viscachas, etc.) and the primates have relatively long gestations. In addition, many marsupials and eutherians have developed a method for prolonging gestation by holding the blastocyst in a state of delayed implantation – or, more correctly, embryonic diapause – for weeks or months on end. We will need to consider each of these exceptions in turn.

The whales, the largest mammals on earth, produce spectacularly large young, weighing several tonnes, after a relatively short gestation length that is only about a year (see Table 2.2). They have achieved this by having an extremely rapid fetal growth rate, and in this they are unique, even amongst the marine mammals (Fig. 2.6). Unlike seals, they do not have to haul out on to dry land to give birth, and so the enormous fetus does not present a particular encumbrance to the mother in her watery environment. It seems likely that the large body size of the parents and offspring is of adaptive significance, since it enables them to withstand the intense cold of the polar seas, where the baleen whales spend much of their

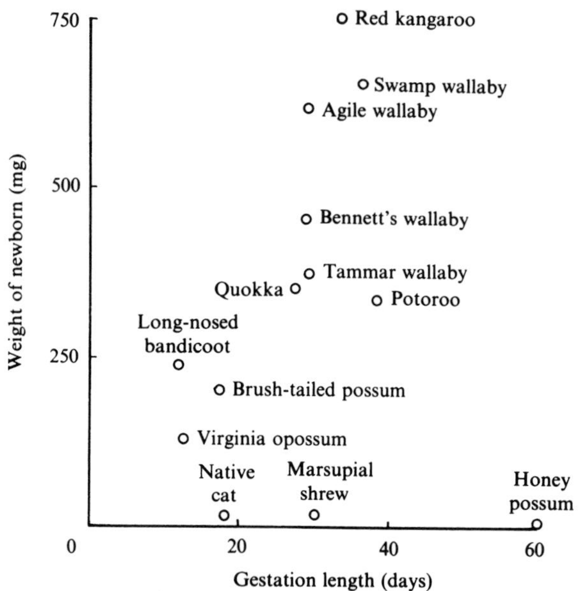

Fig. 2.4. Weights of marsupial newborn young in relation to gestation length. In the case of the macropods, these are gestation lengths in the absence of diapause. The honey possum has an obligatory diapause.

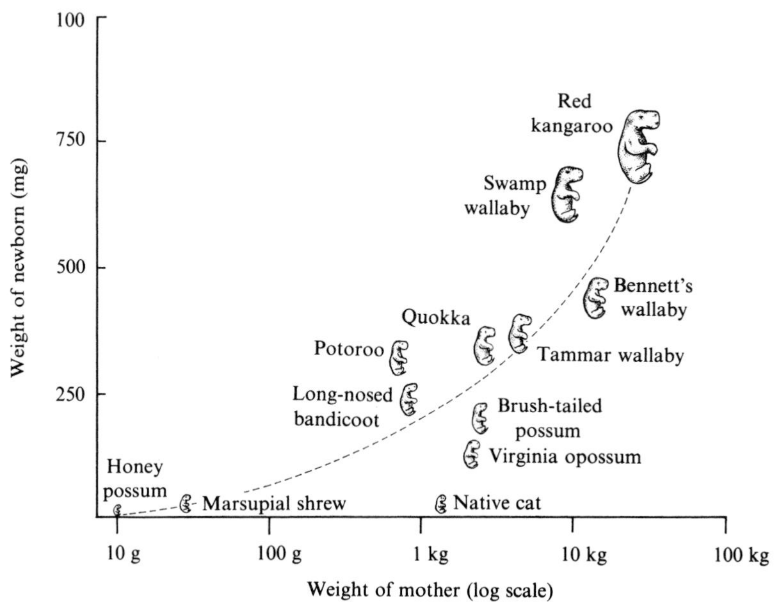

Fig. 2.5. Weights of marsupial newborn young in relation to maternal weights. (From C. H. Tyndale-Biscoe. *Life of Marsupials*. Edward Arnold; London, Fig. 2.6 (1973).)

Table 2.2. *Maternal body weight, gestation length and birth weight in whales*

Species	Maternal body weight (kg)	Gestation length (months)	Birth weight (kg)
Sperm whale *Physeter macrocephalus*	6350–13500	15.5	1054
Fin whale *Balaenoptera physalus*	45000–60000	11.0	1750
Sei whale *Balaenoptera borealis*	15500–18500	11.5	750
Grey whale *Eschrichtius robustus*	13500–35000	13.8	760
Minke whale *Balaenoptera acutorostrata*	5000–8000	10.0	305

(Data from C. Lockyer, personal communication.)

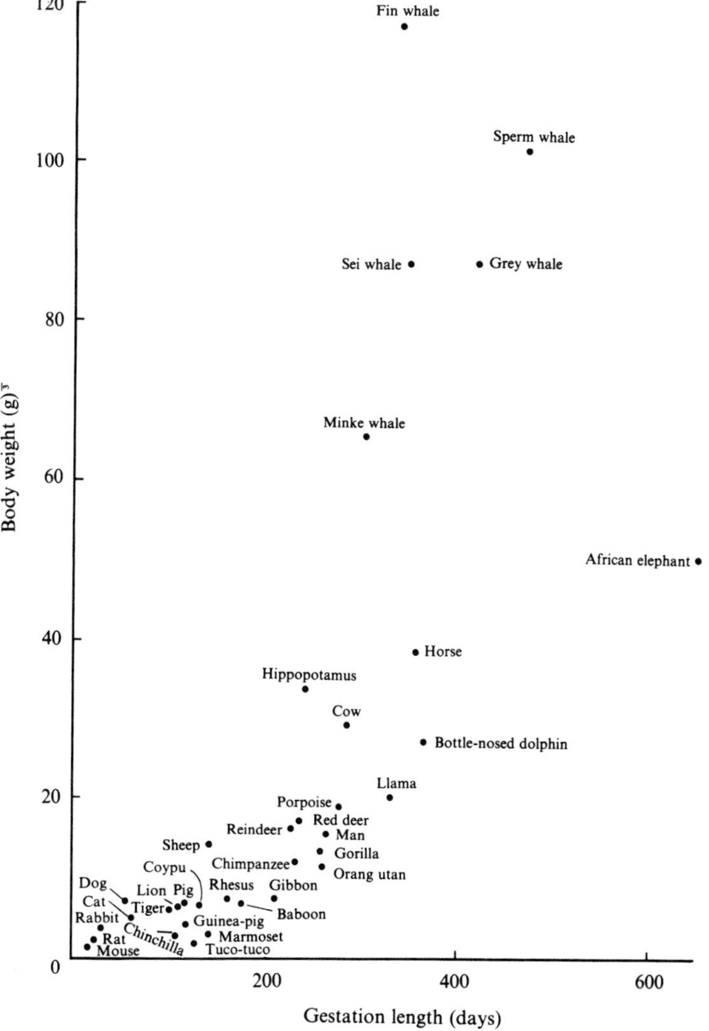

Fig. 2.6. Weights of eutherian newborn young in relation to gestation length. It is known that within a species there is a linear relationship between the cube root of fetal body weight and fetal age throughout most of gestation. Note the extremely rapid growth rates in whales, but not in dolphins or porpoises, and the slow growth rates in hystricomorph rodents, and in higher primates. (Data from G. A. Sacher and E. F. Staffeldt. *Amer. Nat.* **108**, 593–615 (1974); W. Leutenegger. *Folia Primat.* **20**, 280–93 (1973).)

lives feeding off the krill, and the toothed whales also find their prey. It is interesting that the dolphins and porpoises which inhabit more temperate or tropical waters, where large body size would not be so advantageous, have normal fetal growth rates.

The hystricomorph rodents, which are particularly well represented in South America, have spectacularly long gestations for their body size (see Table 2.3). They usually give birth to rather small litters of precocial (advanced) young; it has even been suggested that the guinea-pig is barely a mammal, since the young are so well-developed at birth that they can survive without any of their mother's milk if alternative foods are available. However, it would be wrong to conclude that the prolonged gestation has developed merely as a means of producing precocious young. The guinea-pig in fact has a relatively short gestation by hystricomorph standards; the diminutive tuco-tuco and degu have gestations of 90–130 days, and yet their young are born blind and almost naked, with only the long guard hairs penetrating through the skin. So we must seek another explanation for the spectacularly long gestations in this interesting group of rodents.

The most plausible theory is that hystricomorphs have often chosen to

Table 2.3. *Body weight, gestation length, birth weight and litter size in some hystricomorph rodents*

Species	Body weight (g)	Gestation length (days)	Birth weight (g)	Average litter size
Domestic guinea-pig *Cavia porcellus*	1000	68	100	6
Wild guinea-pig *Cavia apera*	500	61	60	3
Chinchilla *Chinchilla laniger*	500	111	35	2
Tuco-tuco *Ctenomys talarum*	150	130	8	4
Degu *Octodon degus*	250	90	14	5
Coypu *Myocastor coypus*	6000	132	225	5
Capybara *Hydrochoerus hydrochaeris*	30000	150	1500	4
Plains viscacha *Lagostomus maximus*	3000	153	200	2
Canadian porcupine *Erethizon dorsatum*	5000	217	1500	2
Brush-tailed porcupine *Atherurus africanus*	2500	105	150	1

(Data from B. J. Weir, *Symp. Zool. Soc. Lond.* **34**, 265–301 (1974).)

colonize harsh environments at either high altitudes or high latitudes. Under such conditions, where the winters will be long and severe, it would be advantageous for a species like the chinchilla to prolong gestation so as to allow both mating and birth to take place during more clement weather. This probably explains the difference in gestation lengths and birth weights between the Canadian porcupine and its more environmentally favoured African relative. Furthermore, the capybara, the world's largest rodent, weighing in at 30 kg, lives in the tropical swamps and rivers of South America, and has a gestation length that is much nearer to the eutherian norm. Unfortunately, this environmental explanation cannot account for the long gestations of those hystricomorphs that live in subtropical forests, or those that live in underground burrows like the plains viscacha.

The reason why primates should have such long gestation periods relative to their body size, and yet give birth to relatively small offspring, is not immediately apparent (see Fig. 2.6). The small arboreal primates from the tropical rain forests of South America, like the marmosets, tamarins and squirrel monkey, give birth to particularly well-developed young that can successfully cling to their mothers during their aerial acrobatics. Since these species frequently produce twins, there is no way in which the mother herself could hold on to both of them as she swings from branch to branch in search of food. However, an arboreal lifestyle does not necessarily result in a prolonged gestation and precocial young; the orang utan, which is almost exclusively arboreal, has a gestation period that is comparable to that of the terrestrial chimpanzee and gorilla and yet a rather immature offspring.

The one thing that does characterize the primates as a group is their large brain size and greater intelligence. This clearly poses problems, since delivery of a large head is not compatible with the restricted dimensions of the pelvic canal, so that much of the brain growth and skull growth has to occur postnatally. In the higher primates, this encephalization has put the emphasis on neoteny – the retardation of somatic development in fetal and neonatal life – and herein probably lies the reason for the birth of such altricial young after a relatively long gestation in man and the great apes.

Embryonic diapause

Even though it is obviously possible to vary gestation lengths within wide limits by opting for different fetal growth rates, many species seem to have been saddled with gestation lengths that are grossly incompatible with the environment in which they find themselves. The carnivores are a particularly good example; in general they have relatively short gestations and give birth to litters of altricial young in some sort of den or nest. This poses no problems in the tropics, but in more extreme environments, with long, harsh winters, the carnivores were obviously faced with a difficulty. If births were to occur in the springtime, so as to maximize the mother's food supply and the infant's chances of survival, this meant that mating would have to take place in the depths of winter, when it would be better to conserve energy, for example by hibernation. Their solution to this dilemma was to make the embryo 'hibernate'. This was easiest to achieve immediately after mating, when the embryo was still an unattached blastocyst, entirely dependent on uterine secretions for its further growth and development. These uterine secretions are under the control of ovarian steroids, which can be regulated by the hypothalamus and pituitary, and hence by higher centres in the brain. Once implantation has taken place, fetal growth is dependent on the placental transfer of solutes from the maternal circulation, so all central control is lost; thus the only way of slowing down the growth of a fetus is by cooling the whole of the mother's body, as in some hibernating bats (see also Book 2, Chapter 2, Second Edition, for a detailed discussion of embryonic dispause).

Several different groups of carnivores and pinnipeds have opted for embryonic diapause as a key reproductive strategy, but for rather different reasons. Thus, the bears have used it as an adjunct to their winter hibernation; the seals have used it so that mating can take place on the one occasion during the year when all females congregate in one place for pupping, thus making a 1-year gestation period essential; and the mustelids, like the badgers, skunks, stoat, martens and wolverine, have used it to enable them to mate and to give birth during clement weather. However, all these species probably rely on photoperiod to trigger the onset and termination of the diapause, and it is not dependent on lactation (see also Chapter 4). In every case the corpus luteum is held in a state of endocrine quiescence, presumably due to a deficiency of some as yet undefined pituitary luteotrophin, so that uterine secretions are held in abeyance, and the unilaminar blastocyst within its zona pellucida increases in size extremely slowly. The diapause may last as long as 10 months in the case of the wolverine and the European badger. A more spectacular example is the stoat: the female mates in June or July, and gives birth the following April or May, after an extended period of diapause. However, the newborn female offspring reach sexual maturity when only 6–8 weeks of age, before they are weaned, so that copulation occurs whilst they are still in the nest. We do not know who sires this first litter, but it ensures that all the female offspring are already pregnant before dispersing into the wide world.

A possible new addition to the list of carnivores with delayed implantation is the giant panda, weighing 130 kg, which like the bears gives birth to one or two extremely small offspring, weighing 75–120 g, after a gestation of 115–178 days (see Fig. 5.19). Cursed with a carnivore's dentition and digestive tract, whilst having to subsist on a diet of bamboo which even a herbivore might find unpalatable, this species seems doomed to extinction in the wild because of progressive human encroachment into its habitat, and in captivity because of our inability so far to provide it with conditions conducive to normal reproduction.

The small size of bears at birth also fascinated our ancestors, who must occasionally have observed parturition in captive bears in their menageries. They recorded in the Bestiaries how the mother gave birth to a small, amorphous ball of tissue (the amnion-enclosed fetus), and how she sculptured it into her own likeness with her tongue – hence the origin of the phrase 'to lick into shape' (see Fig. 2.7).

Carnivores are clearly something of a special case as far as diapause is concerned, because of their short gestations. This undoubtedly explains why diapause is so widespread in those members of this order that have colonized harsh environments. There is no need to develop diapause as a means of optimizing the time of mating and parturition in species that already have gestation lengths that are well in excess of half a year. Thus the large ungulates of northerly latitudes, like the moose, muskox and caribou or reindeer, with gestations of 7–8 months, are able to cope with the problems posed by their environment. The barren-ground caribou from the extreme north of Canada have improved their situation by undertaking long northerly migrations of up to 800 miles to reach their summer calving and feeding grounds, retreating south again to the forest edges in order to avoid the worst of winter. But problems must begin to arise for the smaller ungulates with shorter gestation periods, who have also chosen to colonize high latitudes or altitudes.

The saiga antelope from the Russian steppes, with its 5-month gestation, is forced to rut in late December in order to calve in the spring, and once again extensive north–south migrations are an essential part of its reproductive strategy. But it is the roe deer, weighing a mere 25 kg, and its larger sub-species, the Siberian roe, that seem to have found the key

Fig. 2.7. 'Ursus the Bear, connected with the word "orsus" (a beginning), is said to get her name because she sculptures her brood with her mouth (ore). For they say that these creatures produce a formless fetus, giving birth to something like a bit of pulp, and this the mother-bear arranges into the proper legs and arms by licking it. This is because of the prematurity of the birth. In short, she pups on the thirtieth day, from whence it comes that a hasty, unformed creation is brought forth.' (From T. H. White. *The Book of Beasts*, p. 45. Being a translation from a Latin Bestiary of the Twelfth Century. Jonathan Cape; London (1954).)

to success by extending the gestation period to 10 months through a 5-month period of embryonic diapause. This has enabled them to extend their range from central and southern Europe and Asia right up to the Arctic Circle in Scandinavia (see Fig. 2.8).

The roe deer is unique in many ways. It is the only ungulate to have developed embryonic diapause, and it also happens to be the species in which the phenomenon was first described by the great German anatomist Theodor Bischoff, as long ago as 1854. The roe deer has also opted for a type of diapause that is anatomically and endocrinologically quite different from that seen in the carnivores and pinnipeds. Following mating at the single oestrus in late July or early August, the fertilized egg develops into a blastocyst, which hatches from its zona pellucida. It then embarks on a 5-month period of arrested development, during which time it still continues to grow, although at a very slow rate; by the end of December it has become a crenated bilaminar structure, several millimetres in diameter, with a layer of endoderm and a clearly defined inner cell mass. The corpus luteum remains fully functional throughout this period of arrest, but there is little endometrial secretory activity, presumably accounting for the slow embryonic growth rate. We still do not understand precisely why endometrial activity should be suspended in this way; perhaps it is due to a low level of ovarian oestrogen secretion. Whatever the reason, at the beginning of January, probably in response to a photoperiodic cue, the endometrium is reactivated, thereby allowing the conceptus to undergo a spectacular elongation prior to developing its first placental attachment to the maternal caruncles. It is interesting that hunters occasionally report shooting a doe in late December or early January with a full-term fetus in the uterus, showing that the delay mechanism sometimes fails to operate. However, there is little chance that a kid born at this time would survive for long. Normally the doe will give birth to one or two kids in late May or early June; there seems to be little

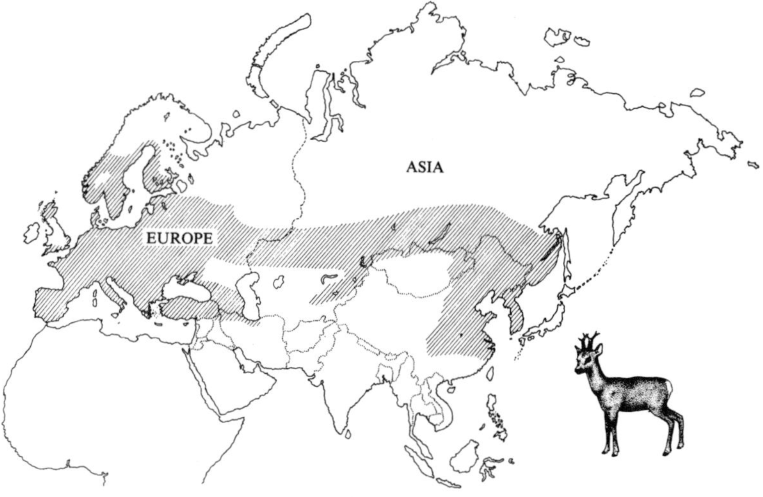

Fig. 2.8. Distribution of roe deer. (From G. K. Whitehead. *Deer of the World*. Constable & Co.; London. Map 14, p. 85 (1972).)

or no embryonic mortality, since each corpus luteum is usually represented by a fetus. We can be certain that diapause is not in any way related to lactation, since does giving birth for the first time, before they have ever lactated, also experience a normal 5-month diapause.

Embryonic diapause also occurs quite commonly in several other mammalian orders, including the rodents (gerbils, voles, rats and mice) insectivores (shrews), chiropterans (bats), edentates (armadillos), and the macropodid marsupials (kangaroos and wallabies). In the rodents and insectivores diapause seems to be used as a device for spacing successive births, rather than for achieving an appropriate timing for mating and parturition, and so it is under lactational rather than photoperiodic control. Since these species have short gestation lengths and can breed repeatedly, often throughout the whole year, there is a very real danger that in the absence of a period of post-partum anoestrus, the birth of a second litter would follow so hard upon the heels of the first that the survival of both sets of young would be compromised. Consider the situation in the laboratory mouse: following a normal gestation of 19 days, the female can become pregnant again within 24 h of parturition. Since her first litter would not normally be weaned for about 25–30 days, it becomes essential to prolong the second pregnancy in order to prevent a premature termination of lactation. Thus, the mouse has developed a lactation-induced period of embryonic diapause; the presence of suckling young can effectively prolong the ensuing gestation for up to 10 days, the precise duration being dependent on the number of young in the litter.

The endocrine mechanism responsible for diapause in the mouse has not been fully elucidated. As in the roe deer, the blastocyst hatches from its zona pellucida, and the corpus luteum appears to be fully functional throughout the delay. The fact that an injection of oestrogen will terminate diapause suggests that the sucking stimulus interferes with pituitary LH secretion, which is probably responsible for the transitory rise in ovarian oestrogen secretion that normally precedes implantation in this species.

A similar story holds true for the laboratory rat, and probably for all the other rodents and insectivores with suckling-induced diapause. But at least one insectivore, the Siberian mole, has made use of seasonal embryonic diapause in the same way as the carnivores, in order to space out mating and birth in a harsh environment. The Siberian mole mates in June or July, and the blastocysts remain in diapause until about the end of March, with births occurring in April and May. It would be interesting to know what determines the length of diapause in this subterranean mammal; can it really be photoperiod, or is temperature involved?

Embryonic diapause is also known to occur in several species of bats, including one of the African fruit-eating bats, *Eidolon helvum*, that actually lives on the equator, and can be seen in enormous numbers hanging like seed pods from the trees that line the streets of Kampala, Uganda. We know too little about the ecology of this species to understand the adaptive significance of its diapause, and nothing is known of the underlying endocrine mechanisms. The Californian leaf-nosed bat *Macrotus californicus* has seemingly done the impossible by opting for an extremely slow rate of embryonic and fetal development, in the absence of any true diapause, so that it has a gestation of 9 months. It would be fascinating to know how this has been achieved, and what advantages it confers.

The delayed implanters *par excellence* are, of course, the large macropodid marsupials from Australia. They can have short gestations of about a month (see Fig. 2.4) and always have an extended extra-uterine development in the pouch lasting up to 10 or 11 months, so the uterus is potentially capable of producing offspring at ten times the rate at which the pouch is able to accommodate them. The macropods have therefore used lactation-induced diapause first and foremost as a means of spacing births to avoid the 'no-room-in-the-pouch' syndrome. There are also a few small marsupials belonging to other orders, like the honey possum *Tarsipes rostratus*, that have diapause. In the honey possum the length of diapause is probably obligatory and neither photoperiodically nor lactationally induced. It seems to be related to prolonging the intervals between births so that emergence of pouch young coincides with the flowering of the different nectar-producing banksias on which these animals feed.

The macropodid marsupials have adopted a variety of different solutions to prevent overcrowding in the pouch, and also to ensure that the young joey eventually leaves it at a time when the food supply is most abundant and the weather is most favourable (see also Book 2, Chapter 2, Second Edition). It must be remembered that, in contrast to eutherian mammals, where it is the timing of birth that is so critical, in marsupials it is the time of pouch emergence.

The simplest strategy to understand is that of the swamp wallaby

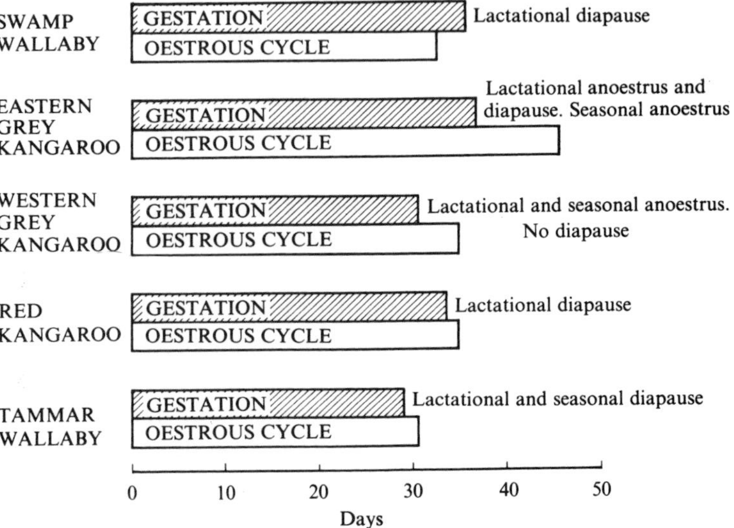

Fig. 2.9. The relationship between gestation length and oestrous cycle length in macropodid marsupials. In all these species except the western grey kangaroo the gestation length is normally extended for up to a year by a long period of embryonic diapause, which is regulated by the suckling activity of the joey in the pouch, and maybe also by daylight length.

Wallabia bicolor, which is highly unusual amongst marsupials in having a gestation that is *longer* than the oestrous cycle (Fig. 2.9). Since oestrus is not suppressed by pregnancy, females conceive again at a pre-partum oestrus, so that for a few days they are carrying two concurrent pregnancies of different gestational ages. This is made possible because of the unusual anatomical arrangement of the marsupial reproductive tract, with its two separate uteri that open independently into the vagina, coupled with a tendency for ovulation to alternate between the two ovaries, so that a new pregnancy can develop in the previously empty uterus on one side of the body, without disturbing the established pregnancy in the uterus on the other side (Fig. 2.10). Following parturition, the sucking activity of the newborn joey, by now firmly attached to a teat, arrests the development of the corpus luteum formed from the new pre-partum ovulation, so that

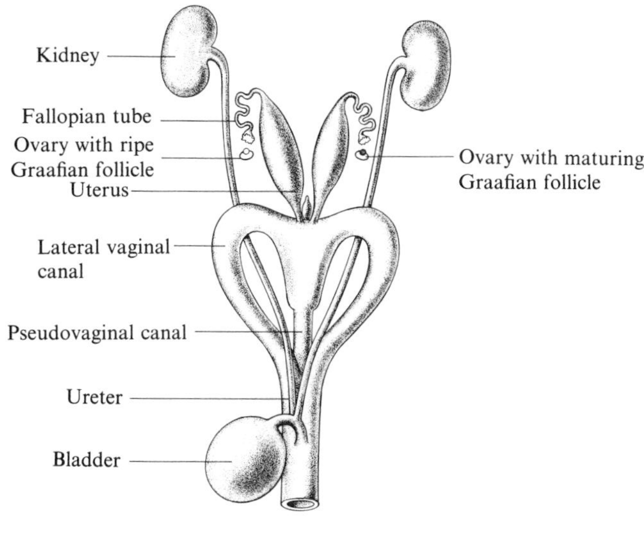

Fig. 2.10. The reproductive tract of a female marsupial. The existence of two separate uteri opening independently into the vagina, and the tendency for ovulation to alternate between the two ovaries, means that pregnancies can follow one another in rapid succession, or even overlap.

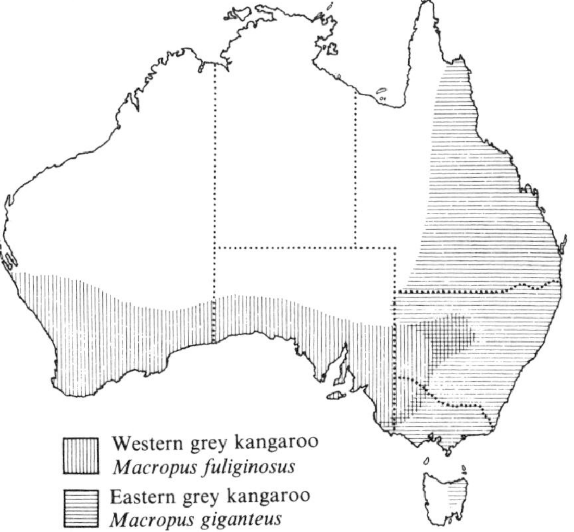

Fig. 2.11. Distribution of the eastern and western grey kangaroo. Although identical in external appearance apart from a darker coat colour (and offensive male smell) in the western grey, they are distinct species. Only the eastern grey exhibits embryonic diapause. (From W. E. Poole. *The Status of Endangered Australian Wildlife*, p. 25. Ed. M. J. Taylor. Royal Zoological Society of South Australia (1978).)

the new blastocyst is held in diapause and will not resume development until the joey eventually stops sucking and vacates the pouch. The female swamp wallaby must surely be unique amongst mammals, since theoretically there is never a day or an hour of its adult life when it is not pregnant!

Next we come to the eastern and western grey kangaroos, once thought to be just sub-species, but now recognized as two entirely distinct species, *Macropus giganteus* and *fuliginosus*; even though their ranges overlap, they do not hybridize with one another, and have rather different reproductive strategies (see Figs. 2.9 and 2.11). The western grey is the only macropod that never has embryonic diapause. In order to prevent overcrowding in the pouch, it uses a seemingly much simpler alternative, adopted by so many of the larger eutherian mammals, namely, lactational anoestrus. The sucking activity of the newborn joey in this case is able to prevent the occurrence of a post-partum oestrus, since the duration of gestation is 4 days *shorter* than the oestrous cycle; the mother will not ovulate again for about 11 months, until the joey stops sucking when it is about to vacate the pouch. The western grey is also a seasonal breeder, and females can come into oestrus only during the summer months, so that the young are ready to leave the pouch only in the summer time of the following year.

The eastern grey kangaroo also has a period of lactational anoestrus after birth, but if food is plentiful the majority of animals will escape from this inhibition and ovulate again during lactation. However, the resultant blastocyst will remain in diapause until the joey leaves the pouch. If food is scarce, then diapause does not occur, and so the reproductive pattern becomes identical to that of the western grey. Like the western grey, the eastern grey is also a seasonal breeder, with matings, and hence pouch emergence a year later, confined to the summer months. The eastern grey has therefore been able to make use of lactational diapause as an 'optional extra', and it would be interesting to know whether this is the factor that has enabled it to colonize a wider range of latitudes than its western cousin.

In the extreme south of Australia, the winters can be quite severe, and it is here that we find species such as Bennett's wallaby *M. rufogriseus*, from Tasmania, and the tammar wallaby *M. eugenii*, from Kangaroo Island, that have used embryonic diapause as yet another way to ensure adequate birth spacing and an appropriate time for vacating the pouch. Following birth of the young in mid-summer, there is an immediate post-partum oestrus, followed by a 4-month period of lactational diapause, when death of the joey results in blastocyst reactivation and a new birth a month later. This lactational diapause then gives way to an 8-month period of seasonal diapause, from early winter to mid summer, when the blastocyst is held in arrest, not because of the suckling stimulus but due to photoperiodic inhibition of the hypothalamus and pituitary in which pineal melatonin secretion plays a central role. This sequence of lactational followed by seasonal diapause ensures that, whatever happens to the joey in the pouch, the blastocyst in diapause can reactivate only at a time that will eventually result in pouch emergence in the spring or summer. Thus these wallabies normally spend 364 days of the year pregnant, although for 11 months of that time the embryo is in diapause.

The final example of the use to which embryonic diapause can be put is the opportunistic breeding system of the red kangaroo, discussed at some length by Bob May and Dan Rubenstein in the preceding chapter. The red kangaroo is an arid-zone species that inhabits the great central deserts of Australia, regions that know no predictable climate or seasons. This animal has therefore become an opportunistic breeder; with a blastocyst in lactational diapause, a joey in the pouch, and a young at foot, it is ideally suited to cope with drought or flood, famine or plenty. It can breed continuously when conditions are favourable, but in times of drought it may have to jettison the young at foot, the pouch young and even the blastocyst in order to ensure its own survival.

All these macropods probably use the same mechanism for maintaining lactational diapause. The evidence at the moment suggests that the elevated levels of prolactin produced in response to suckling hold the corpus luteum in an arrested state of development and inhibit progesterone secretion. This in turn holds uterine secretions in check and thus prevents blastocyst growth and development. The blastocyst can be reactivated experimentally by progesterone injections, which stimulate uterine secretion, or by hypophysectomy, which releases the corpus luteum from the inhibitory effects of prolactin (Fig. 2.12). The corpus luteum is also arrested in seasonal diapause, but whether this is because of a pineally mediated seasonal hyperprolactinaemia, or some other reason, remains to be determined.

We have already seen enough to appreciate that mammals have evolved a spectacular array of mechanisms for altering their gestation lengths, either to enable them to give birth in due season, or to facilitate spacing between successive births, or to allow them to survive in an unpredictable

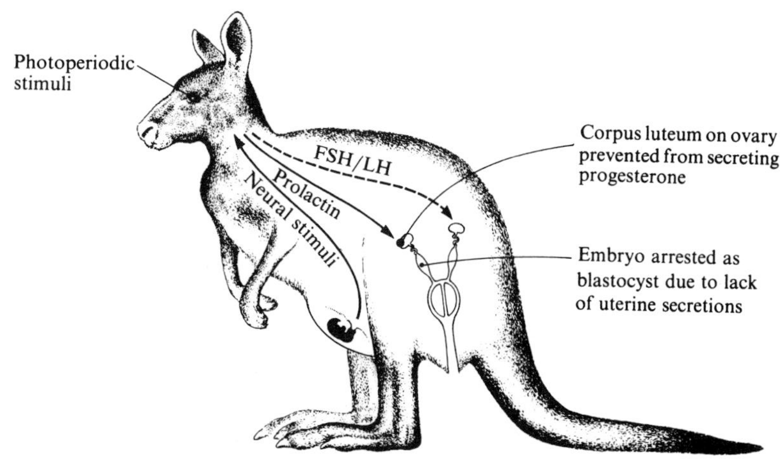

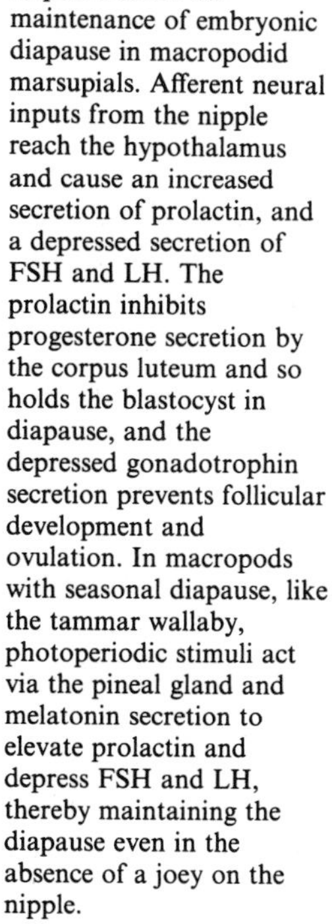
Fig. 2.12. Endocrine mechanisms thought to be responsible for the maintenance of embryonic diapause in macropodid marsupials. Afferent neural inputs from the nipple reach the hypothalamus and cause an increased secretion of prolactin, and a depressed secretion of FSH and LH. The prolactin inhibits progesterone secretion by the corpus luteum and so holds the blastocyst in diapause, and the depressed gonadotrophin secretion prevents follicular development and ovulation. In macropods with seasonal diapause, like the tammar wallaby, photoperiodic stimuli act via the pineal gland and melatonin secretion to elevate prolactin and depress FSH and LH, thereby maintaining the diapause even in the absence of a joey on the nipple.

environment. Fetal growth varies enormously, with the whales winning the prize for the fastest growth rates and largest offspring, the bats for the slowest growth rates, the guinea-pig for the most precocious offspring, and the marsupials for the smallest and most immature newborn young. Each species has evolved a gestation length that is precisely suited to its needs. But, by way of a tailpiece, perhaps the supreme example of adaptation to the environment is seen in a bird, the greater snow goose, which migrates up into the Arctic Circle in the summer to nest. It has been able to cut the incubation time for its eggs from 34 to 24 days, and has accelerated the growth of the young goslings so that they reach 18 times their hatching weight within 3 weeks and are soon fully fledged. In this way, the greater snow goose has been able to exploit the brief 2 months' splendour of an arctic summer for rearing its young, flying south with them to warmer climes before winter closes in once more.

Puberty

Puberty is generally regarded as the 'Great Awakening', the time when the gonads of the male and female first begin to assume their full gametogenic and endocrine responsibilities. However, it must be remembered that the testis of the male is also extremely active endocrinologically during fetal and maybe neonatal life, producing those organizational changes in the reproductive tract and the brain that are essential for normal fertility in later life (see Book 2, Chapter 3, and Book 3, Chapter 6, Second Edition).

Puberty in the male is usually defined as the time at which spermatozoa first appear in the ejaculate. Although in theory this is when the male becomes fertile, in practice what we might term 'behavioural puberty' is not reached for weeks, months or even years after physiological puberty, depending on the species, because the male has to complete his somatic development before he is capable of competing successfully with fully adult male rivals. To take our own species as a case in point, very few 12-year-old boys, who are already producing spermatozoa, would be likely to father children. Wild red deer stags first start to produce spermatozoa when they are a year old, and yet they usually do not become reproductively successful until about the age of 5, when they are sufficiently well-developed physically to defeat their rivals in a contest for a harem of females at the time of the rut.

In contrast to this time-lag between physiological and behavioural puberty in the male, the female usually conceives as soon as she has become fertile, and variations in the age at puberty of the female are therefore of some demographic significance in determining the fertility of the population. Small mammals reach puberty soon after birth, and the all-time record must surely be held by a small South American hystricomorph rodent, the cuis *Galea musteloides*, which can ovulate within 11 days of birth! The timing of puberty in small mammals can be strongly influenced by environmental factors, such as social grouping or pheromonal cues from the male.

It might be thought that all species would have much to gain from reaching puberty as soon as possible, and so we must ask ourselves why puberty is so late in the large *K*-selected mammals. For example, in the whales listed in Table 2.2, puberty does not occur before the age of 6–8 years; is this an inevitable consequence of their greater body size? The timing of puberty is linked in some way to somatic growth; not only must the mother be large enough to deliver her offspring without undue difficulty, but the hormonal changes of pregnancy, and in particular the high oestrogen levels, are likely to cause epiphyseal fusion and hence a cessation of maternal growth in most species, although whales, like elephants, continue to grow throughout their adult lives. Given that these are finite limitations on how early puberty can occur, is there ever a case for the *postponement* of puberty as a reproductive strategy?

The answer appears to be 'yes', at least in the case of our own species. We have already referred to the fact that the increase in size of the brain in higher primates means that much brain growth has to take place in the post-natal period, thereby favouring neoteny. But in addition, the postponement of puberty may have been an enormously important strategy for prolonging the period of childhood dependency, thereby enabling the parents to transmit their vast store of acquired knowledge to their offspring. Unquestionably it has been man's ability to inherit 'acquired' characteristics in this way that has resulted in his supremacy over all other forms of life on earth. It therefore seems no accident that chimpanzees and gorillas in the wild start having menstrual cycles at the age of 8–9 years, and first become pregnant at the age of 9–12, whereas in human hunter–gatherers menarche does not occur until about the age of 16.5 years and because of adolescent sterility they do not get pregnant until they are about 18.

The precise role of puberty in altering the social relationships between parents and their offspring is one of the most interesting and yet least researched aspects of reproductive biology. I was deeply impressed by the results of some experiments that Fiona Guinness and I carried out on some wild red deer on the Isle of Rhum in Scotland. We castrated two stag calves within a few days of birth, and then observed them almost daily for the next 6 years. They were *always* to be found within a few yards of their mothers, whose home range they never left. This was even true during the rut, when the mature stags who were holding the mothers in their harems appeared to accept the castrates just as if they were hinds. The stags could be forgiven for this mistake, since in the absence of testosterone the castrates did not develop any male secondary sexual characteristics, like large body size or antlers, and so they looked almost like hinds. This behaviour is in striking contrast to that of normal stag calves, who remain in their mother's company for the first year of life, but when they reach puberty as yearlings, things begin to happen. Come the rut, they are driven away from their mothers by the harem-holding stags, who will not tolerate their presence. After the rut is over, the yearlings begin to spend more and more time outside their mother's home range, in the company of other stags, and even when they do return, their mothers frequently threaten and reject them. By the time the young stags are 2–3 years old, the chances are that they will have vacated their mother's home range for good, and at the time of the rut they will begin to range widely in the search for hinds of their own. The results of our experiment therefore clearly demonstrate that the rising level of male sex hormones at puberty makes the hind reject her stag calf, and the calf want to leave his mother. This accounts for the fact that it is the young males who are the colonizers in expanding red deer populations; the young females behave like the castrated males and tend to remain in the vicinity of their mothers all their lives. No doubt this male emigration promotes gene flow and prevents inbreeding.

Perhaps we should conclude on a speculative note as far as human puberty is concerned. It seems that improved nutrition in infancy and childhood has been responsible for accelerating the pace of physical development in developed countries, so that menarche in girls now occurs at the age of 12.5–13 years, and puberty in boys is similarly advanced. However, the demands of higher education mean that it is becoming increasingly necessary to postpone childbearing, at least until the late teens. If, as seems likely, human puberty is also associated with hormonally induced changes in behaviour, we have a problem. We now acquire our sexuality, and crave our independence, long before we have the intellectual capacity to cope with either. Trapped between a biological urge and a social taboo, it is not surprising that adolescence places great strains on parents and children alike. The fact that more than 300000 abortions are performed each year on teenagers in the United States reflects not the decadence of youth, but the changing biological nature of man. We have a man-made solution to this man-made problem: contraception.

The birth interval

Probably the most important factor for any species in the regulation of its lifetime fertility is the time interval between successive births. Thus the small mammals, who have done all in their power to accelerate their rate of reproduction, have reduced the birth interval to the very minimum by conceiving again at an oestrus that occurs immediately after parturition. In such species, the birth interval may be only a few days longer than gestation itself.

In general, it seems that the larger the mammal the longer the time elapsed between parturition and the next conception. African elephants, with their 22-month gestation, have birth intervals of 3–4 years, although these can be considerably extended under adverse conditions. Sperm whales, with their 15.5-month gestation, have birth intervals of 4–5 years, although the larger fin and sei whales with their 11.0–11.5-month gestations, have birth intervals of only 2–3 years. The longest birth intervals of all, relative to body size, are seen in the great apes and in human hunter–gatherers; although the duration of gestation in these species is only 8–9 months, birth intervals of 4–5 years are the norm. Before discussing how this is achieved, we should first ask ourselves why it is necessary.

The principal biological consideration to be taken into account when deciding on the optimal birth interval for a species is the time taken for the young to be able to fend for themselves. Some macropodid marsupials have a trick up their sleeves since the mammary gland can produce two different types of milk at the same time, thereby allowing the concurrent support of a young joey permanently attached to one teat, and an older offspring that has vacated the pouch but puts his head in from time to time for a feed from one of the other teats. But for eutherian mammals, where lasting attachment to the teat is not possible, it is essential for one set of offspring to be fully weaned before the next generation is born. The time of weaning is therefore one crucial determinant of the birth interval. No doubt the unusually long period of lactation characteristic of the great apes and human hunter–gatherers is related to their extended period of post-natal brain growth and development, so that they have come to vie with the whales and the elephants for the longest birth intervals.

Another important factor to be taken into account in determining birth intervals is gestation length in relation to the optimal season for giving birth. In species living in harsh environments at high latitudes or altitudes, where births must be confined to 1–2 months of the year to ensure infant survival, a birth interval of 1 year, or multiples thereof, becomes mandatory. If pregnancy is already long, like the 350-day gestation of the wild Przewalski horse from Mongolia, the animal needs to conceive at the 'foal heat' a week or so after parturition, if a 1-year birth interval is to be maintained. But for species with shorter gestation lengths, such as the 5 months characteristic of all wild sheep and goats, a new set of problems arises. Weaning tends to occur relatively early, since infant growth and development have been accelerated to ensure that the offspring are big enough to withstand their first winter. In the absence of prolonged lactation, some other means therefore had to be found of achieving a 1-year birth interval. The solution was to opt for a photoperiodically induced seasonal inhibition of ovarian activity for the greater part of the year, and a precisely timed return of ovulation in the autumn or winter to meet the needs of a spring birth.

We should therefore spend a few moments discussing these two principal ways of prolonging birth intervals in *K*-selected species, namely lactational anovulation and seasonal anovulation. We have already referred to the possible additional effects of non-recurrent embryonic mortality as another way of increasing birth intervals in large mammals.

Lactational anovulation

The simplest fertility-restraining device is to use the sucking stimulus of the infant on the teat to inhibit pituitary gonadotrophin secretion, and hence suppress follicular development and ovulation. The precise endocrine mechanisms responsible for lactational anoestrus, or lactational amenorrhoea as it is called in menstruating primates, have been discussed in some detail in Book 3, Chapter 6, Second Edition.

The key to the inhibitory effect of lactation is a frequent sucking stimulus, maintained throughout the 24 h. Studies of chimpanzees and gorillas in the wild and in captivity show that the mother suckles her infant several times an hour, although each feed lasts for only one or two minutes. The babies sleep with their mothers are night, when frequent suckling almost certainly continues to take place. Recent studies by Mel Konner and his colleagues on the breast feeding behaviour of nomadic !Kung hunter–gatherer women from the Kalahari show that they, too, feed their babies several times an hour for the first 2–3 years of life, and sleep with them at night when suckling continues to take place, even though the mother may not wake up. It seems certain that this high sucking frequency, by day and by night, for 3 or more years, is the main factor responsible for the 4.1-year birth interval found amongst the !Kung. If the infant should die, suckling ceases, and so the mother soon becomes pregnant again.

Figure 2.13 shows what can happen to human birth intervals as the contraceptive effect of breast feeding is eroded by changing social circumstances. The !Kung, who are pro-natalist, do not use modern contraceptives, and practice no form of fertility control such as late marriage, abortion or taboos on intercourse during lactation; nevertheless, they have a mean completed family size of only 4.7, thanks to these prolonged birth intervals. When infant mortality is taken into account, this gives a population doubling time of 300 years.

When we look at the Hutterites, a North American 'back to Nature'

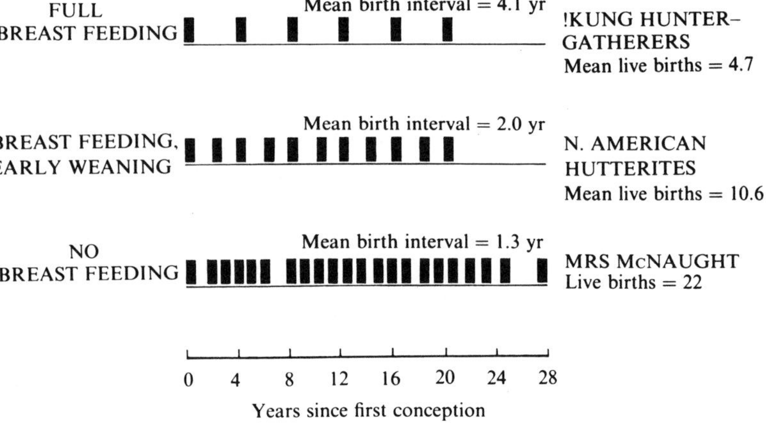

Fig. 2.13. The effect of breast feeding on human birth intervals in the absence of modern contraceptives. (From R. V. Short. Breast feeding. *Sci. Amer.* **250**, 35–41 (1984).)

Anabaptist sect who are also pro-natalist and use no contraceptives, we find that the mean birth interval has dropped to 2 years, and as a result the mean completed family size has jumped up to 10.6. Although the Hutterites also breast feed their babies to begin with, they have started to introduce feeding bottles, powdered milk and 'dummies' or 'comforters' by the third month of life; this must drastically reduce the suckling frequency, and hence diminish the contraceptive effect of breast feeding.

The extreme situation is illustrated by Mrs McNaught, who made the *Guinness Book of Records* for the largest number of single-born children. Following the delivery of each baby, the obstetrician chose to suppress lactation with a large injection of stilboestrol, but he obviously never thought to tell her anything about alternative forms of contraception. As a result, she gave birth to 22 children in the space of 28 years. Had she got married earlier, and suppressed lactation by some means other than stilboestrol, she might even have had one or two more children!

This progressive increase in fertility from the !Kung to the Hutterites to Mrs McNaught probably illustrates in microcosm the changes that have taken place in human fertility within the last few thousand years (see Fig. 2.14). Our hunter–gatherer ancestors, like the !Kung, almost certainly had relatively low fertility and mortality rates (although the mortality would be regarded as unacceptably high by present-day Western standards). This kept human population growth in check for 2 million years or more. Then, as man began to abandon the nomadic hunter–gatherer lifestyle in favour of subsistence agriculture, with the planting of crops, the domestication of animals, and the beginnings of urban development, several changes took place. The increased availability of cereal grains and maybe animal milk provided a ready source of early weaning foods for the baby. The

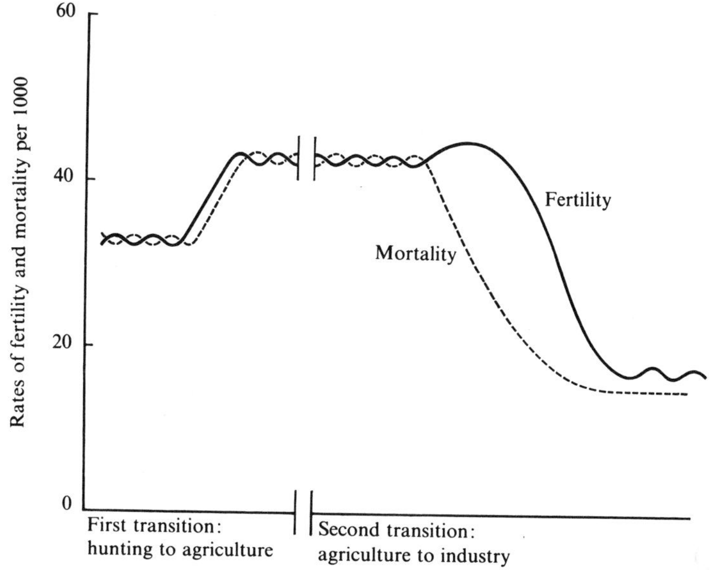

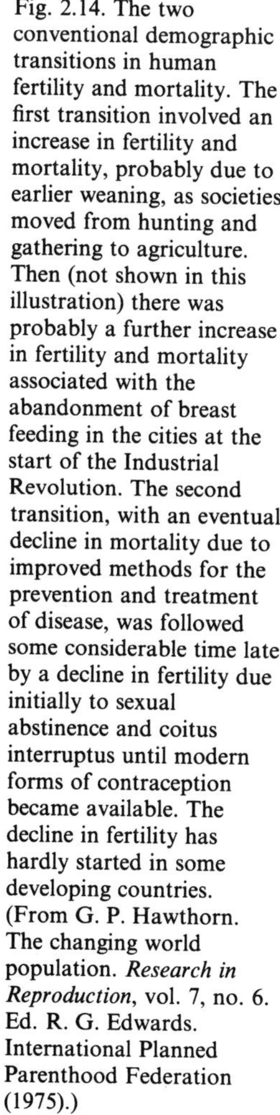
Fig. 2.14. The two conventional demographic transitions in human fertility and mortality. The first transition involved an increase in fertility and mortality, probably due to earlier weaning, as societies moved from hunting and gathering to agriculture. Then (not shown in this illustration) there was probably a further increase in fertility and mortality associated with the abandonment of breast feeding in the cities at the start of the Industrial Revolution. The second transition, with an eventual decline in mortality due to improved methods for the prevention and treatment of disease, was followed some considerable time later by a decline in fertility due initially to sexual abstinence and coitus interruptus until modern forms of contraception became available. The decline in fertility has hardly started in some developing countries. (From G. P. Hawthorn. The changing world population. *Research in Reproduction*, vol. 7, no. 6. Ed. R. G. Edwards. International Planned Parenthood Federation (1975).)

construction of permanent dwellings meant that the baby could now be left in a place of safety, and it was no longer necessary for the mother to carry it with her all the time whilst she worked in the fields, husbanding the crops. Both these factors would have encouraged a reduced suckling frequency and earlier weaning, with a resultant increase in fertility. Infant mortality probably rose at the same time, partly as a result of declining standards of hygiene in the increasingly crowded urban dwellings, but also perhaps because the earlier abandonment of breast feeding deprived the infant of the protective effect of the immunoglobulin A in its mother's milk (see Chapter 6 in this book for further details). The major cause of infant mortality throughout the world is gastroenteritis, and the naturally occurring antibodies in milk provide specific protection against this.

The second great transition in human lifestyles was associated with the industrial revolution, and this probably resulted in a further rise in both fertility and infant mortality. As women entered the workforce, prolonged breastfeeding became impossible, and the children were weaned early on to a diet of 'pap', made of flour, cow's milk and water, all of which were probably heavily contaminated with pathogens. Many women simply abandoned their babies to the care of Foundling Hospitals which sprang up in many of the major cities of Europe. Gastroenteritis killed children in their thousands. In the Dublin Foundling Hospital, 10227 abandoned babies died between the years 1775 and 1796, the mortality rate being 99.6 per cent! In the City of London between 1762 and 1771, there were 162833 recorded births; during the same period, there were 79877 burials of children under the age of 2, and 101454 burials under the age of 5. The nineteenth and twentieth centuries then saw a growing understanding of the causes of disease, and the development of increasingly effective treatments, so that infant mortality began to fall; in developed countries today, infant mortality in the first 12 months of life is down to a figure of about 1 per cent of children born.

It was this decline in mortality in the face of a continuing high level of fertility that led to the explosive growth of human populations. Prolonged breastfeeding had become a forgotten art in developed countries by the nineteenth century, and it has taken us until the middle of the twentieth century to develop alternative forms of contraception so that we can once more hold our fertility in check. Unfortunately, the developing countries are now going through the same demographic transition that took place in Europe in the eighteenth and nineteenth centuries; for them, breastfeeding is still demographically the most important factor regulating fertility. But the young, affluent, urban educated, who are the trendsetters in their communities, are abandoning it in favour of feeding bottles and milk formula. Since only 17 per cent of couples in developing countries today are using modern forms of contraception, there will inevitably be a sustained increase in their population growth rates until such time as the provision of contraceptives more than compensates for the waning influence of breast feeding.

The irony of the situation is that man evolved to be the most *K*-selected of all mammals; but we have by accident eroded the effectiveness of our principal natural birth-spacing mechanism, breast feeding, whilst raising a host of irrational political, economic, social and religious barriers to the deployment of alternative man-made contraceptives.

Seasonal anovulation

The endocrine mechanisms responsible for the regulation of seasonal breeding are discussed by Gerald Lincoln in Book 3, Chapter 3, Second Edition, and the whole subject is dealt with in detail by Brian Follett in Chapter 4 of this volume. However, seasonal breeding is such an extremely important mechanism for regulating the birth interval in mammals living in arctic and temperate regions of the world that we should perhaps cite one or two examples to show how precisely parturition and hence mating has to be timed in order to ensure the maximum survival of the offspring.

Figure 2.15 shows the timing of births in the feral Soay sheep on the island of St Kilda, situated way off the west coast of Scotland at latitude 57 °N. This primitive breed was taken to the island by early man, a thousand or more years ago, and abandoned there to fend for itself. You can see how the timing of births, in mid-April, coincides perfectly with the growth of the pasture in the spring. If births occurred any earlier, there would be no food for the ewes to support the increased energy demands

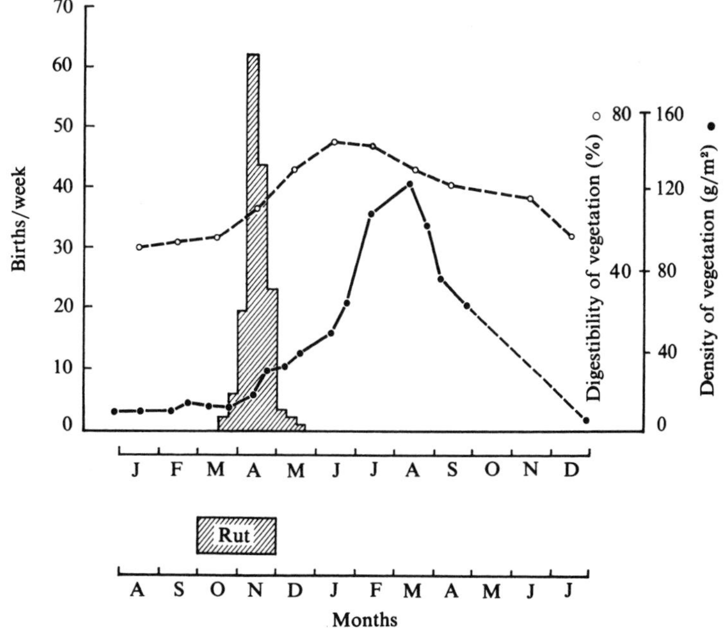

Fig. 2.15. Timing of births (histogram) in feral Soay sheep on the island of St Kilda, off the west coast of Scotland, in relation to the availability and digestibility of herbage on the island. The lower scale, displaced by 150 days (the duration of gestation in Soay sheep) shows the timing of conception as assessed from the time of birth in relation to the duration of the rut. (Data from *Island Survivors: the Ecology of the Soay Sheep of St Kilda.* Ed. P. A. Jewell, C. Milner and J. Morton Boyd. Athlone Press; London (1974).)

of lactation, so the lambs would suffer. If births occurred any later, the lambs might not be sufficiently well grown to survive the rigours of the oncoming winter. In order to achieve this peak of April lambings with a 150-day gestation period, mating must take place in November of the preceding year.

Figure 2.16 shows a slightly different picture for the wild red deer on Rhum, another Scottish island, situated a little further to the south and much closer to the mainland than St Kilda. Spring comes later to Rhum, perhaps due to the influence of the snow-capped mountains that surround the island on three sides. Here the peak of calvings occurs in early June, and once again this appears to be the optimal time for calf survival.

Tim Clutton-Brock and his colleagues have made a detailed study of calf mortality on Rhum; there are many deaths in the months of May to July, and a second peak of mortality in March and April, at the end of the calf's first winter. If this summer mortality is analysed in detail, it can be shown that about 28 per cent of the earliest-born calves will die, as compared to only 12 per cent of those born at the beginning of June; the mortality rises again to about 28 per cent for those born at the end of the month. One of the principal factors influencing this summer mortality is the birth weight of the calf, which presumably reflects the bodily condition of the hind in late gestation. Any hind that calves early will be giving birth before her bodily condition has had a chance to improve with the advent of spring. Any hind that calves late will have come into oestrus late in the preceding autumn, presumably because she was in poor condition, and she may not have been able to regain condition during the winter.

The winter mortality of calves shows a slightly different picture. Birth weight is an irrelevant factor, but birth date is most important. For

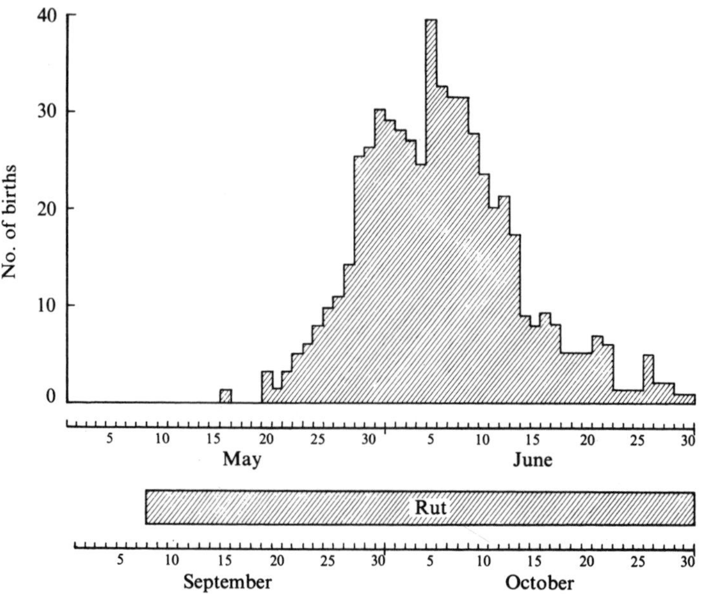

Fig. 2.16. Timing of births in wild red deer on the island of Rhum, off the west coast of Scotland. The lower scale, displaced by 233 days (the duration of gestation in red deer) shows the timing of conception as assessed from the time of birth in relation to the duration of the rut. (From G. A. Lincoln and F. E. Guinness. The sexual significance of the rut. *J. Reprod. Fert.* Suppl. **19**, 475–89, text-fig. 1 (1973).)

example, 19.4 per cent of calves born after the median birth date for their year died, as compared to only 10.3 per cent of calves born before the median date.

We have been talking in terms of an overall mortality in the first year of life of up to 47 per cent for calves born only a few weeks after the optimal time, as opposed to a 22 per cent mortality for those born at the most favoured time. All this must relate back to the time of conception during the rut, 233 days before calving; we can begin to appreciate just how critical this timing is for the survival of the species. Thus, it really is true that the sins of the fathers (and mothers) are visited upon the children of the succeeding generation.

The role of the male

It will not have escaped the reader's attention that so far in this chapter we have been concerned almost exclusively with female reproductive mechanisms. This is no accident. The combined demands of gestation and lactation mean that female mammals have a far greater energy investment in reproduction than males. Thus it is the female that is always the limiting reproductive resource, and so it is hardly surprising that it is the females that show the greatest variety in their reproductive mechanisms.

The prime concern of all female mammals must be to ensure an adequate food supply for themselves and their dependent offspring at all times, and it is this that probably determines the mating strategy for the species. For example, where food is abundant the whole year round, as in some tropical regions, the female can tolerate the continual presence of the male, and so the mating system may be one of continuous rather than seasonal breeding, and lifelong monogamy. Good examples of this are the gibbons from the tropical rainforests of Malaysia, and the marmosets and tamarins from the tropical rainforests of South America. But monogamy is comparatively rare amongst mammals, and polygynous or promiscuous mating systems are the norm. In many regions of the world, food is in short supply for part of the year, and at such times the male becomes an unwelcome competitor who must be excluded from the female's home range. The necessity for seasonal births in such an environment dictates the time of mating, which is the only occasion during the year when it is necessary for the males and females to associate with one another; for the rest of the time, they may behave almost as if they were entirely separate species, living separate lives in different habitats. Red deer are a good case in point, with the stags spending 10 months of the year in bachelor groups, well away from the hinds. At the time of the rut in the autumn, the dominant stags will try to obtain a harem of hinds, usually on favoured lowland grazings where the home ranges of several hinds overlap. However, for all their intensive herding behaviour, the stags will not succeed in moving any individual hind out of her own home range, an area of a few hectares in which she spends her entire life. Edwin Landseer's famous painting of 'The

Monarch of the Glen' may look the epitome of male supremacy, but in fact the glen will be 'hind ground', and the stag will have been forced to migrate there in order to seek out female company at the time of the rut.

Barry Keverne discusses in Chapter 5 the way in which the mating system has had a profound influence on the phenotypic appearance of the male as a result of sexual selection. In monogamous species, the male and female are of similar body size and appearance, whereas in polygynous species the males are usually larger and heavier than the females, and have developed offensive weapons, such as horns, antlers or large canine teeth. All these male secondary sexual characteristics are produced under the influence of testosterone, and it is testosterone that is responsible for the seasonal changes in male behaviour at rutting time.

In species where the female shows seasonal reproductive quiescence, this is often reflected in the male, but the effects are usually neither as pronounced, nor as prolonged. In every case it seems to be the female rather than the male that determines the time of the mating season. For example, ewes come into oestrus for the first time in September or October, and if not mated will continue to have regular 16-day oestrous cycles until the following February or March, whereas rams are fertile throughout the whole year, although their testosterone levels, libido and semen quality are depressed during the summer months when the ewes are in the depths of anoestrus. Red deer hinds first come into oestrus in October, and if not mated will have 18-day oestrous cycles until February, whereas the stags are fertile from September to March.

We have recently become interested in a specific aspect of male sexual selection, which is best described as gonadal selection; it appears that the mating system has repercussions not only on the somatic phenotype of the

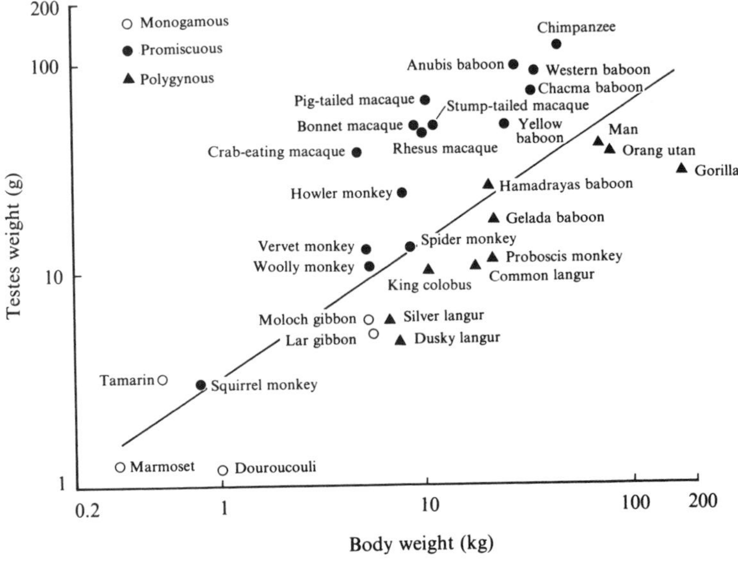

Fig. 2.17. Testicular weight in relation to body weight in primates. Species above the line have promiscuous, multi-male mating systems, whereas those below the line are either monogamous or polygynous. (Data from A. H. Harcourt, P. H. Harvey, S. G. Larson and R. V. Short. *Nature*, **293**, 55–7 (1981).)

male, but also on the size of his gonads. This can be clearly seen in the case of the primates (Fig. 2.17). If we plot testicular weight against body weight for a whole range of primates, the ones with the largest testes are those with the highest copulatory frequencies, and promiscuous mating systems, like the baboons, macaques and chimpanzees. In monogamous species, like the marmoset, or in polygynous species like the gorilla where copulation may occur as infrequently as once or twice a year (see Book 8, Chapter 1, First Edition), and even in man himself, the testes are extremely small relative to body weight.

The reason for this striking difference is simple enough to understand. The spermatogenic capacity of the testes is determined ultimately by the volume of seminiferous tubular tissue; if more spermatozoa are required, the only way of producing them is to increase the volume of the production plant, because the rate of production per gramme of testicular tissue is fixed. Since the seminiferous tubules make up over 90 per cent of the volume of the testis, changes in testicular volume provide an accurate reflection of changing spermatogenic capacity. In a promiscuous mating system, each male potentially has access to a larger number of females, so that copulatory frequency is increased. In addition, if more than one male copulates with a female at a given oestrus, gamete selection will favour the male with the largest number of spermatozoa in his ejaculate, and that male is most likely to father the offspring. Large testes also occur in seasonally breeding polygynous species, like rams or stags, where all the copulations have to be confined to a few brief weeks during the rut.

It is difficult to avoid the conclusion that the male mammal is henpecked, since his reproductive strategy is largely determined by and dependent upon the needs of the female. Although she may have little opportunity for exercising any active choice in selection of a mate in most polygynous mating systems, intra-male competition will ensure that only the best male becomes her suitor.

Lifespan

There can be no doubt that small mammals have short lifespans, but why should this be so? One might imagine that the optimal trade-off for any animal would be the longest possible lifespan with the shortest possible generation time. However, in an unstable and constantly fluctuating environment a small animal may soon outlive its usefulness. The survival of the species could perhaps be best served by rapidly reshuffling the genetic pack of cards at meiosis and fertilization, thereby producing a new deal of offspring, some of whom might be better suited to cope with the environmental change, whilst at the same time bringing about the early mortality of the obsolescent parents. But there are also penalties associated with constantly tracking short-term unpredictable environmental changes; the investor who opts for small profits and quick returns can never hope to make long-term capital gains and dominate the market. Large animals, with their greater bodily reserves, may be more able to weather the temporary vicissitudes of life; they can survive the lean years, to profit from the years of plenty that follow. Clearly we need to look upon lifespan as another adaptive characteristic that has also been subjected to intense selection pressures during the course of evolution.

Is a 100-year-old mouse a biological impossibility simply because all small animals 'burn themselves out' after 1–2 years as a result of their high metabolic rate? In dogs, it is the toy breeds that live longest, and the big breeds like wolfhounds and great danes have the shortest lifespans. Birds, with their elevated body temperatures, have particularly high metabolic rates, and yet some species, such as ravens and parrots, can live to a great age. Perhaps we could get some valuable clues about the general phenomenon of ageing through studying the ecological benefits conferred upon certain mammals, birds and reptiles by their greater longevity.

We already have some fragmentary evidence in a variety of species to suggest that lifespans are in part hormonally determined, and hence linked to the reproductive system. The effects are most apparent in males. Studies in a Kansas institution for the mentally retarded, where until quite recently some men were castrated to make them more manageable, showed that the castrates lived on average 13.6 years longer than intact men in the same institution (see Fig. 2.18), although there was no obvious medical explanation for this life-sparing effect. Normal men between the ages of 35 and 54 in the USA are known to be three times as likely to die from heart disease as women, and there could be an underlying hormonal

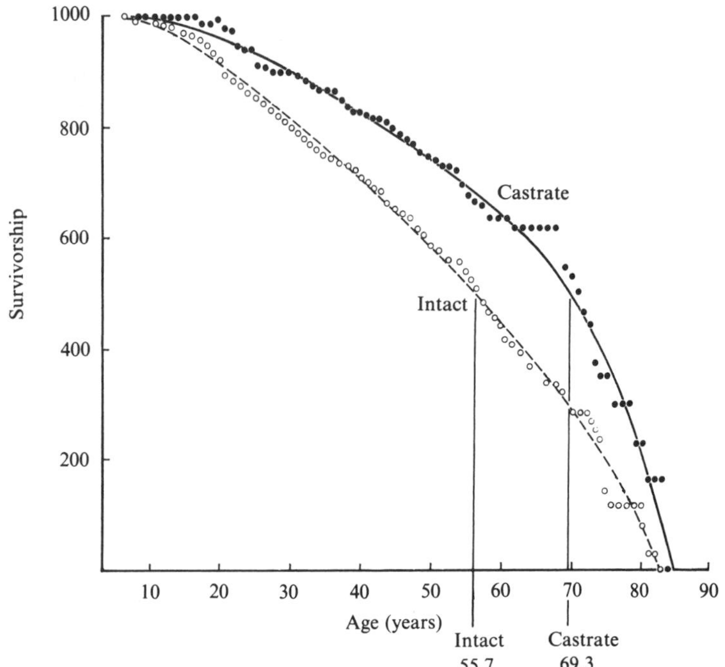

Fig. 2.18. Survivorship curves of 735 intact and 297 castrated Caucasian men living in an institution for the mentally retarded in Kansas, USA. The median survivorship of the castrates was 13.6 years longer than the intact men. Men castrated at 8–14 years of age were also longer lived than those castrated at 20–39 years of age. (Data from J. B. Hamilton and G. E. Mestler. *J. Gerontol.* **24**, 395–411 (1969).)

explanation for this also. Peter Jewell and his colleagues have studied natural mortality in the Soay sheep of St Kilda, and have shown that after puberty the mortality of males is much greater than that of females for every year of their lives (see Book 6, Chapter 3, First Edition); however, if the rams are castrated soon after birth, this preferential male mortality disappears. The most plausible explanation in this case is that the intact males lose a great deal of bodily condition during the autumn rut, and therefore may not be able to survive the ensuing winter. The castrates, who have little or no urge to rut, are therefore at a considerable advantage.

The most dramatic and best understood effect of androgens on male mortality is seen in a marsupial shrew, *Antechinus stuartii*, and two other closely related species, *A. swainsonii* and *A. flavipes*. Ian McDonald and his colleagues in Melbourne have shown that the abrupt, total mortality of all the males in wild populations about 2–3 weeks after the single, intensive annual mating period in late winter is due to a sudden rise in free plasma corticosteroid concentrations; this produces gastrointestinal ulceration and haemorrhage and a depressed immune response, so that the animals ultimately succumb to a variety of endemic parasites and microorganisms. McDonald has been able to show that this mortality-inducing rise in free corticosteroids is androgen-dependent; testosterone concentrations rise abruptly just before the mating season, to produce frenetic territorial aggression between rival males, increased adrenocortical activity and adrenal hypertrophy. However, the elevated testosterone levels also cause a severe suppression of the corticosteroid binding capacity of the plasma, presumably due to impaired hepatic synthesis of corticosteroid-binding globulin; as a result, there is an enormous increase in the amount of free corticosteroid available to the tissues, resulting in eventual death. If males are castrated prior to the mating period and released into the wild, they can survive for another year. Females also survive to breed again for a second year, since they always have an adequate corticosteroid-binding capacity to cope with the increased adrenocortical activity during the mating season. The most plausible ecological explanation for this bizarre mortality is that it removes the highly aggressive males from the population just before the weaning of the young in the spring, thereby preventing competition for the limited supply of insect food.

There is much less evidence to suggest that sex hormones are involved in regulating the lifespan of females. However, it should be remembered that in our own species there are many documented hazards of nulliparity, such as an increased incidence of cancer of the breast, ovaries and uterus (see Book 3, Chapter 6, Second Edition). This could have been of some adaptive significance, since it would have helped to remove infertile individuals from the population.

The topic of reproductive senescence is discussed in more detail by Cyril Adams in Chapter 7, but it is interesting in the present context to consider the possible biological significance of the menopause as it relates to the general problem of lifetime fertility in long-lived, *K*-selected species. If women remained fertile throughout their adult lives, perhaps they would simply produce too many children for the habitat to support, notwithstanding the 4-year birth intervals as a result of prolonged breastfeeding. The fact that the menopause is confined to the female also seems significant if it is to be viewed as primarily a fertility-regulating device. Unfortunately, we do not know enough about other long-lived, large mammals; elephants can occasionally live to about the age of 60, and there is some evidence to suggest that very old female elephants may sometimes experience reproductive failure. Fin whales have been recorded as living as long as 96 years, and sei whales as long as 74; with birth intervals of only 2–3 years, and puberty at 6–11 years of age, this would potentially give 6–8 decades of reproductive life with 3 or more offspring per decade. Unfortunately, we do not know whether cetaceans experience a menopause.

We must also ask ourselves whether there are any evolutionary advantages to be gained from having non-reproducing individuals in the community, where they may be competing for scarce resources with the fertile population. In human societies, post-menopausal women can continue to play an invaluable role in the upbringing of their children and grandchildren. However, it must be remembered that any diseases that occur in the post-reproductive years will be largely beyond the reach of genetic selection, and maybe this accounts in part for the increased incidence of diseases such as breast cancer after the menopause.

The whole concept of lifespan as a reproductive strategy has hitherto received scant attention, and unfortunately we lack an adequate data base from which to formulate hypotheses. Few people have had the foresight or the opportunity to study the factors contributing to lifetime reproductive success for a species in its natural habitat, and yet this is what natural selection and evolution is all about.

The natural regulation of animal populations

It seems appropriate to conclude this chapter with some consideration of the mechanisms mammals can use to bring their rates of population growth into equilibrium with their environment.

Perhaps this is the very thing that the small, *r*-selected mammals have never been able to achieve. They may be doomed to periodic decimation by famine, flood or frost, so that their lives are constantly at the mercy of their environment. In the laboratory, it is possible to set up experiments to demonstrate the effect of one particular environmental constraint on population growth. In one such study, John Calhoun placed five pairs of wild rats in a 10000-ft^2 enclosure with free access to unlimited food and water, and he studied their behaviour for the next 2 years. At the end of that time, when the population could have been in excess of 50000 had each individual continued to reproduce maximally, there were only 171 survivors. This appeared to be mainly due to a complete social and behavioural disorganization of the community; few of the newborn young survived because of highly abnormal maternal behaviour. Eventually, it was only the dominant animals in the population that managed to reproduce successfully. Similar results have also been obtained with mouse populations kept in like circumstances. However, in Nature it seems more likely that a shortage of food, rather than space, would be a limiting factor, and we do not know precisely how a rat or mouse population, given unlimited space, would react to the constraint of food.

The population biologists of the 1960s like John Christian and Dennis Chitty, were heavily influenced by the ideas of Hans Selye about the effects of stress on the adrenal cortex, and they saw 'adrenal exhaustion' as a mechanism for the regulation of small mammal populations. Few people would now accept that this has any role to play in population regulation, although it cannot be denied that the adrenal glands appear to play a key role in regulating the numbers of the marsupial shrew *Antechinus*, at least as far as the males are concerned.

Graeme Caughley and Charles Krebs from Canberra have recently suggested that the mechanisms underlying the natural regulation of animal populations can be regarded as either extrinsic, such as shortage of space or food, or intrinsic, such as physiological or behavioural adaptations within the animal. They have proposed that population growth in the smaller herbivores (less than 30 kg body weight) is likely to be regulated in the main by intrinsic factors, and in larger herbivores by extrinsic ones.

Although it is true that the Calhoun experiments demonstrate very clearly how an intrinsic behavioural adaptation can regulate rodent populations, this is only in response to an extrinsic stimulus, namely a shortage of space. Brian Follett and Barry Keverne have discussed in succeeding chapters a whole variety of intrinsic physiological mechanisms such as pheromones, and behavioural mechanisms such as social grouping effects, that can regulate the fertility of small mammals. But we must not forget that many small mammals are equally prone to extrinsic regulation through factors like the photoperiodic control of seasonal breeding.

As far as the large mammals are concerned, extrinsic factors such as food supply undoubtedly play a major role in regulating fertility. Consider the situation in wild red deer. We have already discussed the enormous importance of birth weight, time of birth and subsequent growth rate on calf survival. Calf mortality is probably the single most important factor determining population growth. To this can be added effects of nutrition on age at puberty and on subsequent birth intervals. Well-grown red deer calves reach puberty during their second year of life, but if their growth is stunted by a shortage of food, they may not reach puberty until a year later. Hinds that are on poor grazings will produce less milk than normal, so that their calves suckle more frequently; this results in a delayed return to oestrus, which will have adverse repercussions on the survival of the next calf when it is born the following summer. The energy demands of lactation may be so great that the hind loses condition badly during the summer months, and fails to come into oestrus at all during the autumn, so that she calves only in alternate years. Thus it is not difficult to see how a combination of extrinsic factors such as food supply and daylight length can regulate the growth rate of the red deer population through a subtle interplay with intrinsic physiological regulatory mechanisms.

As far as human populations are concerned, extrinsic factors have also been of major significance in times past in regulating infant mortality and hence determining the rate of population growth. With improved nutrition and the conquest of disease, these extrinsic controls have been relaxed, and unfortunately we have at the same time chosen to tamper with the principal intrinsic regulatory mechanism, namely the birth-spacing effect of breast feeding. It has been this simultaneous reduction in mortality and stimulation of fertility that has led to the explosive growth of the world's population.

In this chapter we have considered all creatures, great and small, and they have presented us with a dazzling array of adaptations of the reproductive system, designed during the course of evolution to meet a variety of different environmental demands. The plasticity of the system is truly amazing. And we have barely scratched the surface, for the great majority of the world's mammals have yet to be investigated; sadly, we will have exterminated some of the most spectacular ones before we have had a chance to study them. Even for the species that we know best, like our laboratory and domesticated animals, we still have only a hazy idea of the environment in which their reproductive systems evolved. Until we understand that we cannot begin to make sense out of this seemingly endless reproductive diversity. I am reminded of Henry Beston's memorable sentiments about animals:

> We patronise them for their incompleteness, for their tragic fate of having taken form so far below ourselves, and therein we err, and greatly err. For the animal shall not be measured by man. In a world older and more complete than ours they move finished and complete, gifted with extensions of the senses we have lost or never attained, living by voices we shall never hear. They are not brethren, they are not underlings, they are other Nations, caught with ourselves in the net of life and time, fellow prisoners of the splendour and travail of the earth.

Suggested further reading

Are big mammals simply little mammals writ large? G. Caughley and C. J. Krebs. *Oecologia*, **59**, 7–17 (1983).

Aspects of the reproductive endocrinology of the female giant panda (*Ailuropoda melanoleuca*) in captivity, with special reference to the detection of ovulation and pregnancy. J. K. Hodges, D. J. Bevan, M. Celma, J. P. Hearn, D. M. Jones, D. G. Kleiman, J. A. Knight and H. D. M. Moore. *Journal of Zoology*, **203**, 253–267.

Endocrine changes in Dasyurid marsupials with different mortality patterns. I. R. McDonald, A. K. Lee, A. J. Bradley and K. A. Than. *General and Comparative Endocrinology*, **44**, 292–301 (1981).

Embryonic diapause in mammals. Ed. A. P. F. Flint, M. B. Renfree and B. J. Weir. *Journal of Reproduction and Fertility*, Supplement **29** (1981).

Reproductive characteristics of hystricomorph rodents. B. J. Weir. *Symposia of the Zoological Society of London*, **34**, 265–301 (1974).

The environment and reproduction in mammals and birds. Ed. J. S. Perry and I. W. Rowlands. *Journal of Reproduction and Fertility*, Supplement **19** (1973).

Patterns of Mammalian Reproduction (2nd ed.). S. A. Asdell. Cornell University Press; Ithaca, New York (1965).

Red Deer: Behaviour and Ecology of Two Sexes. T. H. Clutton-Brock, F. E. Guinness and S. D. Albon. Edinburgh University Press (1982).

The Mammalian Radiations. J. F. Eisenberg. University of Chicago Press (1981).

Marshall's Physiology of Reproduction, Vol. 1. Reproductive Cycles of Vertebrates. (4th ed.). Ed. G. E. Lamming. Churchill Livingstone; London (1984).

Island Survivors: the Ecology of the Soay Sheep of St Kilda. P. A. Jewell, C. Milner and J. Morton Boyd. Athlone Press; London (1974).

3

Genetics and reproduction

R. B. LAND

The pattern of reproduction of an individual is determined by its inheritance, but reproduction is also the means by which the individual acquires and transmits its inheritance. The genes passed from one generation to another ensure that like mates with like and that offspring resemble their parents. Studies of the genetic control of reproduction therefore tend to revolve around the assessment and interpretation of the extent to which reproductive traits are determined genetically and hence of the similarities among relatives. Such studies can help our understanding of reproduction in three ways.

First we might look for common elements in the pattern of reproduction of members of a particular genetic group and so aid understanding by the development of unifying concepts. For example, 'What are the common characteristics of reproduction among mammals?' is a genetic question and the extent to which mammals resemble each other can be appreciated only by comparison with other classes and groups from which they differ.

Then we might use genetic variation in the search to identify causative functional relationships among steps in physiological pathways. Similar concentrations of, say, luteinizing hormone (LH) in the plasma of individuals of genetic groups giving birth to different numbers of young would indicate that variation in LH is unlikely to be involved in variation in litter size.

Also, we might use knowledge of the physiological expression of genetic variation to recognize the variation itself. It would be of considerable significance to agriculture to be able to overcome the fact that the expression of many reproductive traits is limited by both the sex and the age of the individual; this would, for example, enable us to measure the genetic merit of a young male for litter size.

The study of genetic variation is a particularly valuable, if often neglected, part of the study of the physiology of reproduction. Knowledge of the physiology of variation not only aids understanding but also indicates the opportunity for change through selection, and hence, in an agricultural context, for improvement. It is the aim of this chapter to illustrate the extent of genetic variation and its contribution to our understanding of reproduction.

Litter size

The most obvious expression of reproductive activity and the most readily quantified is the number of young born in a litter. This varies among species – a difference that is by definition genetic – and also within species, where variation as a whole may arise from both genetic and environmental sources.

Variation among species

The differences among eutherian mammals in the individual weight and number of their young at birth are enormous. Mice, shrews and voles at one extreme give birth to individuals weighing of the order of 1 g, while at the other extreme newborn whales weigh 1 tonne. On the one hand we might ask how can a single pattern of reproduction cope with such variation, but on the other we might ask why, if it can cope with such variation, it cannot cope with more. An examination of the relationship between birth weight, maternal weight and litter size helps us to answer these questions.

May and Rubenstein have already drawn attention in Chapter 1 to the tendency for the litter size in eutherian mammals to decrease as maternal weight increases (Figs. 1.8 and 1.9), and the one simple characteristic that stays relatively constant is the weight of individual new-born eutherian mammals as a percentage of adult maternal weight; this varies around an average of 5 per cent. What seems to have happened in the course of evolution is that newborn eutherian mammals that weigh less than 3 per cent of adult weight tend to be too small to survive, and those that weigh more than 10 per cent tend to be too large to survive parturition unscathed.

If the relationships are studied in more depth, however, another constant

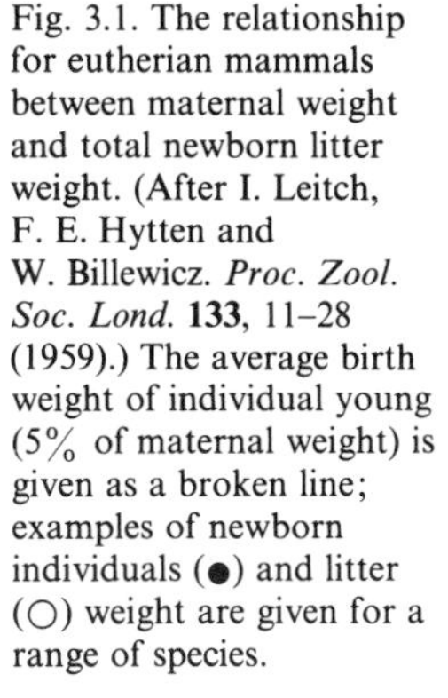

Fig. 3.1. The relationship for eutherian mammals between maternal weight and total newborn litter weight. (After I. Leitch, F. E. Hytten and W. Billewicz. *Proc. Zool. Soc. Lond.* **133**, 11–28 (1959).) The average birth weight of individual young (5% of maternal weight) is given as a broken line; examples of newborn individuals (●) and litter (○) weight are given for a range of species.

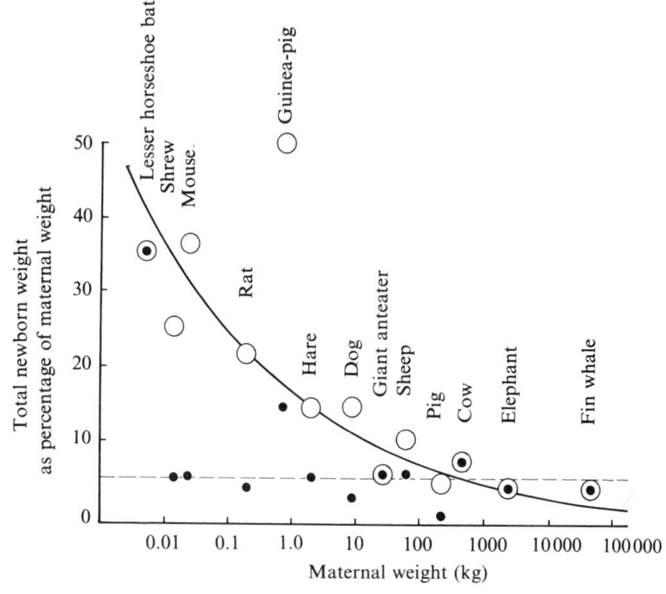

emerges: the weight of the litter is related to maternal weight, not simply, but logarithmically. The log of litter weight in grammes is equal to (0.8 times log maternal weight) − 0.33. Converting from logs, litter weight would be expected to be equal to half of maternal weight to the power 0.8. The other relationships now begin to fall into place. With total litter weight related to a power of maternal weight less than one, it has to decline as an absolute percentage of maternal weight as maternal weight increases. We already know, however, that the birth weight of an individual eutherian mammal does not fall much below 3 per cent of maternal weight, and so if 50 per cent of log maternal weight to the power 0.8 is less than 3 per cent of maternal weight, the species would not survive. The reproductive characteristics of eutherian mammals have put an upper limit on their size range (Fig. 3.1).

With litter size increasing as a proportion of maternal weight as maternal weight decreases, and with an upper limit on the relative birth weight of individual young, natural selection had the opportunity to favour multiple births in small mammals and, as we said at the beginning, litter size tends to be greater in small mammals than in large ones. Presumably, from a reproductive viewpoint, the size of mammals could decline below that of existing species; but then factors such as food intake and the control of body temperature would tend to play a limiting role.

The essential *mammalian* components of the physiology of reproduction have therefore led to a tendency for eutherian mammals to follow a particular physical pattern of reproduction. This tendency is not absolute; just as individual species deviate from the common pattern in the physiological basis of their reproduction, so too do they deviate from the common physical pattern. Polar bears, for example, produce single young which weigh only 0.3 per cent as much as their mothers; bats at the other end of the scale give birth to individual young that may be one-third the weight of their mothers – 100 times as much. It is, however, the generalities that enable the deviations or exceptions to be recognized. Pigs, for example, give birth to large litters of lighter individual weight than would be expected. Once deviations are recognized, associated problems can be appreciated or opportunities exploited.

Variation within species

Variation in litter size among individuals within a species has been studied extensively in laboratory animals, both as a component of fitness and as a model for domestic mammals, and in domestic mammals themselves with a view to agricultural improvement.

Laboratory animals. Genetic variation in litter size was first studied as a correlated response to selection, that is as a characteristic of lines selected for other traits such as body weight. MacArthur was one of the first to illustrate such changes when he reported in 1964 that females of lines of mice weighing 28 and 14 g as a result of differential selection for body weight had litter sizes of eleven and six young, respectively. Much of the study of this problem has since been led by Douglas Falconer of Edinburgh and Eric Bradford of Davis, California.

The first of Falconer's now classical series of selection experiments in mice was designed to study the direct and correlated responses to selection for body weight. Again, litter size increased in a line selected for large body weight, and vice versa. The body weight and litter size of the high line increased to 30 g and eight young, compared with 13 g and four young in the low and 20 g and seven young in the unselected control line. The range of lines studied by Bradford, however, shows that such a correlated change is not always found. One line (G), selected for weight gain between 3 and 6 weeks of age, weighed nearly twice as much at 6 weeks of age as did mice of the control strain, yet had a similar litter size (eight young). These studies demonstrated the presence of variation in the genes controlling litter size in these populations, and showed that there is a general but not absolute tendency for high litter size to be associated with high body weight, and vice versa. The next series of studies was based upon selection for litter size itself.

Again both Falconer and Bradford demonstrated that the regular choice of females with extremely high or extremely low litter size as parents of the next generation would lead to progressive changes in the litter sizes of the selected lines. The results of Falconer's first experiment are illustrated in Fig. 3.2. At the end of 30 consecutive generations of selection, the litter size of the high line was nine, the low six, and the unselected control 7.5. Most of the change, however, took place in the first 20 generations. The lines are then said to have reached the 'limit' of the response to selection, or to have 'plateaued'.

The characteristics of the response to selection allow us to compare the 'genetics' of litter size with the 'genetics' of other traits. Does litter size respond more or less rapidly to selection, and are the limits of the response

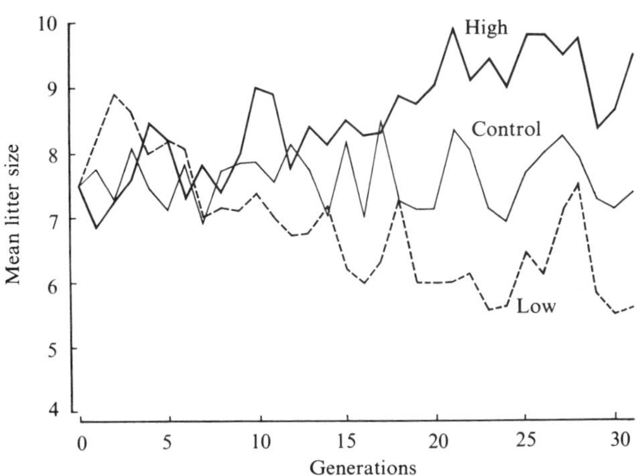

Fig. 3.2. Progress of selection for high and low litter size in mice, together with the litter size of an unselected control line. (After D. S. Falconer. Quantitatively different responses to selection in oposite directions. In *Statistical Genetics and Plant Breeding*, pp. 487–90. Ed. W. D. Hanson and H. F. Robinson. Nat. Acad. Sci. – National Research Council; Washington, publ. no. 982 (1963).)

greater or less? To answer these questions we can introduce some measures of the characteristic of the response.

In absolute terms, the response of litter size measured as the divergence between the high and low lines was 3.2 mice, or 42 per cent of the mean of the selected lines and the control, somewhat less than the 75 per cent divergence between the body-weight lines when expressed relative to the base population. When we compare the responses relative to the standard deviation of the base population, the difference is much greater; the response in litter size was 1.7 times the standard deviation, whereas that of the growth lines was 9 times.

The response of litter size to selection was therefore less than that of body weight, but to interpret this we need to think of what selection was actually doing. A trait responds to selection if selected individuals carry and hence transmit genes slightly different from those of the unselected group. It responds if variation in the form or phenotype of an individual is caused by variation in the genes it inherits from its parents; it is the variation that is important. The basic form of an individual is determined by its genes, but the genes may allow some scope for variation; they may allow variation to arise from the effects of the environment or they may contribute directly to variation themselves by differing among individuals.

To assess these sources of variation the concept of *heritability* has been developed; this is a measure of the proportion of the variation in a character that arises from differences in inheritance. Together with an absolute measure of the overall level of variation, this enables us to describe the genetic control of traits. If there is no variation, the trait is determined absolutely by inheritance, and all the genes controlling that trait are the same. If there is variation, the heritability tells us the extent to which that variation arises from genetic sources. A human example might help make this clear: the brown versus blue colouration of the eye is determined by inheritance; language or accent at the other extreme is determined by environment.

Now, we can use the data from the selection experiments to estimate the heritabilities of litter size and body growth. Falconer and his colleagues were able to select. There was therefore variation. We have said that heritability is an estimate of the proportion of that variation that is transmitted genetically, and so we can see that the proportion of the superiority of selected parents that is transmitted to their offspring gives us a convenient way to measure heritability. To do so, Falconer developed the technique of regressing the performance of each generation on the accumulated selection that had preceded that generation; this gives estimates of about 22 per cent for the heritability of litter size, compared to nearer 40 per cent for body weight. There is, nevertheless, more relative overall variation for litter size than for body weight in a normal population; the coefficient of variation in litter size is about 25 per cent, compared to 10 per cent for body weight.

What does this mean? Litter size, in comparison to body weight, is a variable character, but less of that variation is genetic. The effect of segregation at loci affecting the trait is less and there is less opportunity to change the mean by genetic selection. In evolutionary terms, this difference is related to the closeness of the characteristic to fitness. It may be considered 'important' for animals to have a particular litter size, and so genetic variation is less. Nevertheless, the extent of the change must not be underestimated. By persistently collecting genes for large litter size in one population and for small litter size in another, the females in the high line give birth to 50 per cent more young per litter than those of the low line, as can be seen in Fig. 3.2.

The lines stopped responding when the genetic variation in the base population became exhausted. Had a larger base been sampled originally, the number of segregating loci would have been expected to be greater and the ultimate response larger, but this would also have applied to body growth. Again, as a generalization it is possible to introduce extra variation by crossing and extend the response, but there is always the problem that it may take some time to overcome the initial drop on crossing.

Correlated responses to selection for reproductive characteristics tell us more about the way in which reproduction is controlled genetically. The associations among components of reproduction will be the subject of later sections, but one characteristic that is confirmed is the tendency for high litter size to be associated with high body weight and vice versa, exactly the opposite to the relationships observed between species. How do the forces of natural selection lead to such different results to those of artificial selection? Principally because natural selection applies to the overall characteristics of an individual, not just a single trait. The lines selected for litter size invariably have lower fertility, and life-time reproduction rate is often less.

The within-population studies therefore tell us that there are genes segregating in populations which affect body weight and litter size in the same direction. Among species, the genes with common effects on the two traits that are fixed are genes that have reciprocal effects on the two traits, and it is these differences that prescribe the differences among species. One might argue that the positive-correlation genes are still segregating; because they do not contribute to fitness and because they have not been subjected to the pressures of natural selection, both '+ +' and '− −' genes would be equally unfavourable.

Do the between-species relationships have any implications for the possibilities of selection within populations? The one characteristic that remains relatively constant within a species is the duration of pregnancy, yet between species larger animals have longer gestation lengths. Selection for large body weight would therefore be considered to be limited by the ability of young to grow to more than about 3 per cent of their adult weight before the end of gestation. The positive relationships between body weight and gestation length could reasonably be expected, and Fig. 3.3 shows that the relationship is linear with log scales. Equally, the constancy of gestation length within a species might reasonably be expected because by definition a species is an interbreeding population, and we know that the synchrony of mother and young is a prerequisite to successful pregnancy. St Clair Taylor gives the lovely example of dogs. It would be predicted from the regression in Fig. 3.3 that if the different breeds of dog were different species, the gestation length of genetically large ones would be twice that of small ones; the observed difference is only one-tenth of that predicted.

Laboratory animals have given us considerable information about the inheritance of litter size. In addition to the results of selection, there is also an extensive literature on the effects of inbreeding and cross-breeding, and of single genes.

As a generalization, inbreeding depresses reproductive performance, with the implication that heterozygosity is 'good' for mammalian reproduction. The existence of fecund inbred lines shows that such a relationship is not absolute, but such lines are the exception rather than the rule. Of 100 inbred lines of mice maintained by brother–sister matings, only about five would be expected to survive. The reproductive performance of the others would be so low as to lead to their extinction.

In contrast to inbreeding depression, reproductive traits tend to show *heterosis* or hybrid vigour, i.e. the performance of crosses exceeds that of their parents. This decline and rise is illustrated in Table 3.1, which shows the average litter size of 30 strains of mice during inbreeding, and the crosses among them. It is relevant to note that on average the crosses at the end of the programme produced 0.4 more young than the average of the lines that went into them. It is results of this nature as well as theory that led to the argument that inbreeding may be a particularly effective way of 'selecting out' unfavourable genes. Only those 'good enough' to

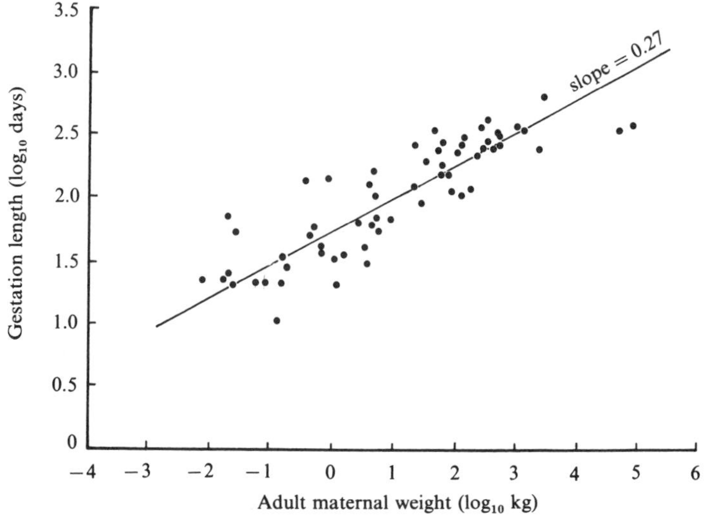

Fig. 3.3. Inter-species relationship of gestation length to mature (maternal) body weight. (After St C. S. Taylor. Use of genetic size scaling in evaluation of animal growth. *Proc. Int. Growth Symp.*, Guelph, 7–8 August 1982. *J. Anim. Sci.* (in press).)

be compatible with viability as homozygotes could survive. Once the unfavourable genes were eliminated, the removal of inbreeding by crossing would yield the useful population. In general, however, this approach has not been found to be appropriate in domestic mammals; the lost opportunities for positive selection of favourable alleles have outweighed the advantages, and direct selection is a more effective procedure.

Before leaving inbreeding and crossing, it is relevant to note that one must be very careful about the use of the word heterosis. In a biological context, heterosis is a measure of the extent to which the average of all crosses exceeds the average of all parents. In an agricultural context, heterosis is useful only when the performance of a cross exceeds that of the best available parent.

Domestic animals. Genetic variation in the litter size of domestic mammals is best recognized by variation among existing breeds. This variation is most pronounced in sheep, intermediate in pigs and least in cattle. The variation in sheep is at least threefold; breeds such as the Romanov and the Finnish Landrace (Fig. 3.4*b*) produce an average of 2.5 lambs to a single mating, while at the other extreme the Tasmanian Merino (Fig. 3.4*a*) produces an average of only 0.75. While the Tasmanian Merino is an example of one extreme, the Booroola Merino is an example of another: its litter size is similar to that of the Finnish Landrace and Romanov, and we shall discuss it later as an example of a single gene effect on reproductive performance.

Within populations, genetic variation in litter size had tended to be presumed to be low, so low in fact that it had been considered not feasible to change the litter size of domestic mammals by genetic selection. The

Table 3.1. *The average litter size of 30 lines of mice during inbreeding by mating of full brothers and sisters for three generations and of 60 crosses amongst them*

Mating pattern	Inbreeding coefficient* Parents	Offspring	Litter size
Outbred	0	0	8.1
Full sibs	0	0.250	6.7
	0.250	0.375	5.8
	0.375	0.500	5.7
Outcross	0.500	0	6.2
	0	0	8.5

* The inbreeding coefficient expresses the probability that both alleles at a locus are of common descent.

(After R. C. Roberts. The effects on litter size of crossing lines of mice inbred without selection. *Genet. Res., Camb.* **1**, 239–52 (1960).)

heritability can be measured in unselected populations by comparing the variation among offspring of different males to that among the offspring of single males, by the use of established Analysis of Variance procedures.

Such calculations give estimates of the order of 10 per cent or less in sheep, cattle and pigs, and it was generally concluded that we could not

Fig. 3.4. (*a*) Tasmanian Merino ewe with one lamb and (*b*) Finnish Landrace with three. (Courtesy of J. D. Barker.)

improve litter size by selection within populations. Helen Turner of Australia has been the principal research worker to dispel this myth for sheep. One very clear example of the success of selection within a population is that of L. R. Wallace in New Zealand. After 20 years, ewes in the line selected for high litter size produced 50 per cent more lambs than those of the control line: 1.49 versus 1.08 per female mated. The performances of the two selected lines are illustrated in Fig. 3.5. As Helen Turner says in her 1969 review, 'Success has thus followed selection...in spite of earlier pessimistic heritability results'. Techniques of genetic selection within populations are now being used as a means to improve litter size in sheep with conspicuous success. In one New Zealand 'Group Breeding Scheme', for example, after only a few years of selection rams were estimated to have a genetic merit of 28 lambs per 100 ewes above that of unselected controls. One reason for the contrast between pessimistic expectations and subsequent success might be the preoccupation of early investigators with heritability as the sole index of the likely response to selection. Many have since drawn attention to the fact that the variability of the trait also affects the rate of response; the greater the standard deviation, the greater the superiority of, say, the top 10 per cent of the population above the mean, and we have already seen for laboratory animals that the coefficient of variation of litter size is 2.5 times that of body weight. The same applies to domestic animals. If the coefficient of variation doubles, the selection response doubles, just as it would if the heritability were twice as great.

Programmes with pigs have been less successful and less persistent. Attempts with cattle are only just beginning, led by Laurie Piper and Bernie Bindon of Turner's school. They have shown that the trait is repeatable and that the incidence of twinning in the daughters of selected cows and bulls is greater than that in the general population (7 per cent versus close to zero) – a promising start to an exciting future.

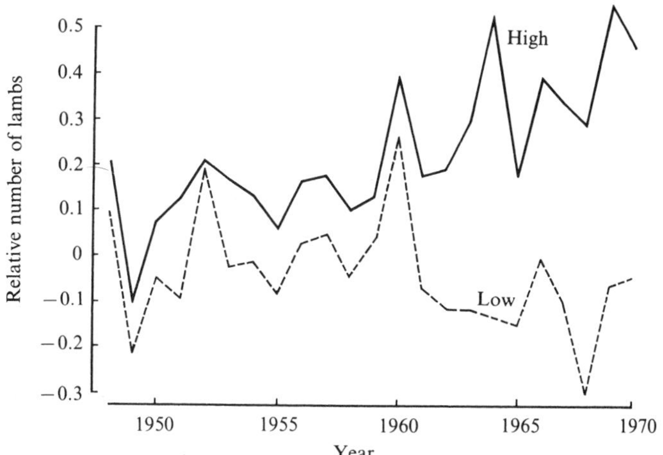

Fig. 3.5. Progress of selection for (High) or against (Low) twinning in Romney sheep. The performance of each line is characterized as the number of lambs born per ewe mated relative to unselected controls. (After J. N. Clarke. Current levels of performance in the Ruakura fertility flock of Romney sheep. *Proc. N.Z. Soc. Anim. Prod.* **32**, 99–111 (1972).)

As with laboratory animals, the reproductive performance of domestic animals is depressed with inbreeding and shows heterosis on crossing. In pigs, for example, of 146 Large White lines started by brother–sister matings, only 18 (12 per cent) survived beyond three generations, to give an inbreeding coefficient of 40 per cent. On average, litter size at birth declines by roughly 0.2 piglets per 10 per cent increase in inbreeding of the dam, and by roughly 0.1 piglet per 10 per cent increase in inbreeding of the piglet. It can be seen from the example in Table 3.2 that 10 per cent inbreeding could have a serious effect on the commercial merit of a pig population: the combined effects of inbreeding of the dam and the litter would reduce litter size by about half a pig, or 5 per cent.

The importance of separating 'biological' heterosis – superiority over the mid parent – and 'agricultural' heterosis – superiority over the best parent – was indicated in the section on laboratory animals. Another potentially serious complication is the environment, for if one parent does not thrive in a particular environment, it is possible for a crossbred to show distinct heterosis for reproduction, or indeed other traits, simply because it inherits sufficient genes from the other parent to enable it to resist the adverse effects of that environment. An example is the crossing of Border Leicester and Merino sheep in Australia, where crossbreds have been found to produce 50 per cent more lambs than the mean of their parents. The Border Leicester breed, however, tends not to thrive in Australia, so it is quite possible that the 'heterosis' arose partly from the depressed performance of this breed. If so, the heterosis would not be expected in all environments, and indeed would be expected to decline as the environment changed to one in which both parents could express their potential fully. In practical terms, the 'origins' of the heterosis do not matter as it is still improvement for that environment, but it does affect our ability to generalize and to predict what might happen elsewhere.

Female fertility

The genetics of fertility defined strictly as the ability to conceive are in general much less well understood than the genetics of the size of the litter, given that an individual conceives. The exceptions to this are the knowledge of genetic variation in timing of puberty, the duration of fertility in seasonal breeders (which includes most mammals) and where fertility has

Table 3.2. *Examples of the effects of a 10 per cent increase in inbreeding on the litter size of pigs*

Inbred generation	Mean number born alive	Mean number alive at 56 days
Litter	−0.13	−0.34
Dam	−0.23	−0.23

changed as a correlated response to selection for other characteristics. There is little general evidence for variation in fertility among mammalian species.

Correlated changes

With most extensive selection having been conducted in laboratory animals, the most obvious changes are also to be found in these species. If we go back to studies of the mouse-growth selection lines of Falconer, we find that both high and low lines suffered from reduced fertility. After 30 or so generations, 18 per cent of the matings were sterile in both selected lines, compared to none in the controls. Bradford reports that the fertility of one of his growth lines of mice fell to 42 per cent at one time, whereas that of the controls was always above 90 per cent and often 100 per cent. These two examples are sufficient to demonstrate a very clear decline in fertility as a correlated response to selection for another trait, and hence genetic variation for fertility must have existed in the base population. The worry in agricultural practice is that the same might happen in domestic animals selected for high levels of production. High-yielding dairy cows, for example, have a tendency to reduced fertility. Low conception at a young age is now a cause for concern in very lean, very rapidly growing strains of pigs. Even if animal breeders do not consider it economically worth while to incorporate measures of fertility into their selection indices with a view to the improvement of fertility, they might do so to prevent the deterioration of fertility in lines highly selected for other traits.

Bradford has looked further at the question of the fertility of lines of mice selected for other traits; while some have maintained their fertility, none has developed higher levels of fertility than that of the unselected controls. Selection for other traits, then, has increased the frequency of 'sterility alleles', and not decreased them.

With fertility as such an evident component of fitness, it is interesting to speculate why natural selection has not led to the elimination of genes for sterility. Possibly the duration of oestrus is a compromise. The shorter the oestrus, the greater the likelihood of mating at a precise time relative to ovulation, and hence of conception. Conversely, the greater the duration of oestrus, the greater the likelihood of mating. Certainly there is genetic variation for the duration of oestrus in, for example, sheep, in which breeds may differ twofold.

Seasonality

Seasonal breeding is discussed at length in other chapters, and they cover genetic variation in terms of differences among species. Some species ovulate and are fertile when daylength is increasing, and the vole and ferret are well studied examples; others ovulate, mate and conceive when daylength is decreasing and the sheep and goat are species that do this in high latitudes. The sheep, however, also illustrates genetic variation within a species, and shows both the flexibility of natural selection and the potential of genetic variation, which aid us in the search to understand physiological mechanisms, a topic we shall return to later.

In the sheep, studies in northern Europe indicate that the period of ovarian cyclicity of indigenous breeds continues from October to March. Sheep therefore tend to mate in October and November and lamb after a 5-month gestation in March and April, as grass grows in the northern spring. As we move south towards the Sahara, however, the limits to grass growth become dictated by water rather than by temperature, so that in North Africa the breeding season is partly reversed, the incidence of ovulation being highest from June to October and at a minimum in February, in breeds such as the Timhadite of Morocco. At intermediate latitudes, in France, for example, indigenous breeds resume ovarian cyclicity in July (Fig. 3.6). If these differences were maintained when the three breeds were placed in the same environment and hence found to be genetic, they would provide excellent material to test hypotheses regarding the control of seasonality. Does sensitivity to gonadal steroid feedback (Book 3, Chapters 1 and 3) change at the same time of year in each breed, or at the same time relative to the onset and end of ovarian cyclicity?

The heritability of the time of onset of the breeding season has been estimated in sheep by studying the similarities between relatives in a

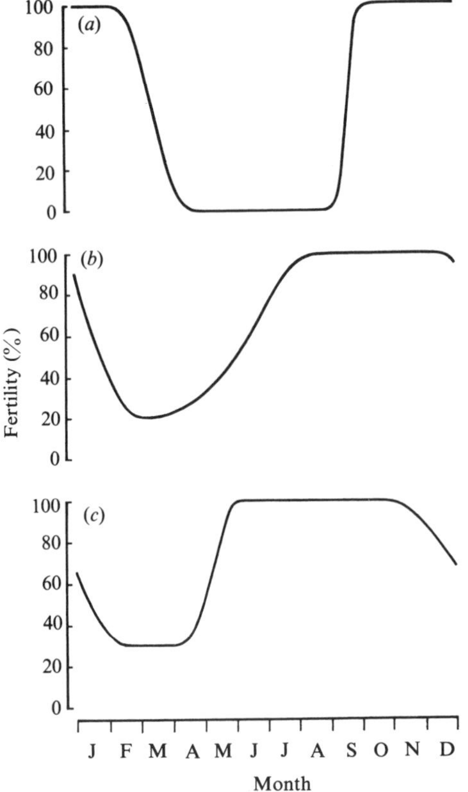

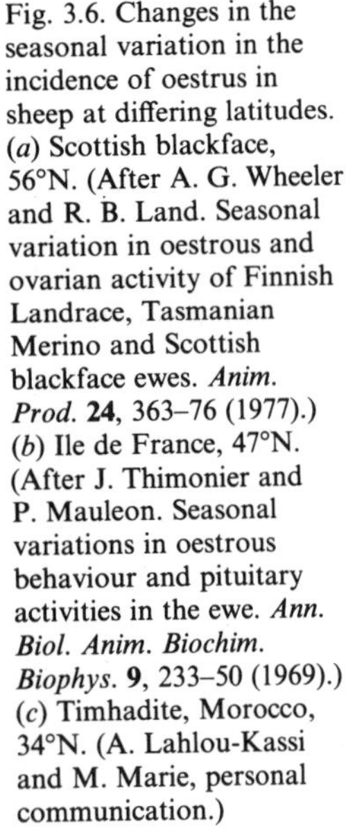
Fig. 3.6. Changes in the seasonal variation in the incidence of oestrus in sheep at differing latitudes. (*a*) Scottish blackface, 56°N. (After A. G. Wheeler and R. B. Land. Seasonal variation in oestrous and ovarian activity of Finnish Landrace, Tasmanian Merino and Scottish blackface ewes. *Anim. Prod.* **24**, 363–76 (1977).) (*b*) Ile de France, 47°N. (After J. Thimonier and P. Mauleon. Seasonal variations in oestrous behaviour and pituitary activities in the ewe. *Ann. Biol. Anim. Biochim. Biophys.* **9**, 233–50 (1969).) (*c*) Timhadite, Morocco, 34°N. (A. Lahlou-Kassi and M. Marie, personal communication.)

pedigree Welsh Mountain flock. This was found to be 0.35, much higher than the 2–5 per cent heritability of conception at oestrus.

Humans versus domestic species

Species differences in seasonality are one exception to the generalization that there is little variation in fertility among species. The apparently low fertility of man may be another exception, but without 'oestrus' and the consequent synchronization of mating and ovulation, it is particularly difficult to compare the fertility of man with other species. Roger Short, in his review, quotes Henry, the French demographer who estimated fertility as the probability of producing a full-term infant per menstrual cycle during which intercourse occurred. Henry defined this as fecundability and found from the study of village records that it was about 25 per cent for newly married women in French rural communities in the eighteenth century; it is presumed that intercourse rates were high. Subsequent studies of other social groups that do not practice contraception have confirmed this estimate. The relationship with frequency of intercourse is, however, shown by a drop in fecundability when the frequency of intercourse declines from at least four times per week to three to four or one to two times per week. Even with the luxury of frequent insemination close to the time of ovulation, only 50 per cent or so of cattle would be expected to give birth. Would cattle be any more fertile than man if they were inseminated at random with respect to the ovarian cycle, and only two to three times per week? Behavioural characteristics might contribute to man's lower fertility.

Ovulation rate and embryo mortality

Many geneticists have turned their attention to variation in the number of eggs shed by the ovary at the time of ovulation, in attempts to understand variation in litter size. Again, there is information from many sources: variation among lines selected for other traits, among lines selected directly, from studies within populations and from variation among species. Ovulation rate has appealed to the geneticist for several reasons, not least of which is that it is easy to score relative to other physiological characteristics, it can be easily defined without ambiguity and it can be seen to set an upper limit to litter size. Ovulation has been investigated as a component of fitness and as a means to increase the reproductive rate of domestic animals.

Variation among populations

Variation in ovulation rate among species is in most cases closely related to variation in litter size. Embryo survival by contrast tends to be remarkably constant in different species, 70 per cent or so of eggs shed being represented as young at birth in most eutherian mammals. Large mammals, therefore, rarely shed more than one egg, while small ones shed up to 20. (One remarkable exception is in the group of hystricomorph rodents, where the plains viscacha may shed several hundred eggs at each ovulation but give birth to only one or two young.)

Lines selected for other traits show the presence of genetic variation in ovulation rate or at least show that it was present in the population from which they were selected. The comparison of lines of laboratory animals selected for a variety of traits, but particularly body growth and litter size, shows that genetic variation in embryo survival contributes to genetic variation in litter size within populations. One of the earliest series of experiments to show this was again one performed by Falconer and his colleagues. Surprisingly, selection for high and low litter size in mice led to increases in ovulation rate in both groups. At the time it was tempting to try to interpret such a change in terms of the genetic variation in ovulation rate and embryo mortality, and the genetic correlations between them and litter size. The dangers of such interpretations are now appreciated and the possibility that this could have arisen by chance as a result of 'genetic drift' is recognized. The powers of artificial selection are illustrated by the fact that, even as a correlated trait, Falconer's high- and low-body-weight lines differed twofold in ovulation rate. In the litter size lines, embryo survival in the low one had fallen to 56 per cent, compared to 81 per cent in the controls. Again, Bradford found embryo mortality to have dropped to 55 per cent in a line selected for low litter size, compared to 75 per cent in the control population, and with hardly any change in ovulation rate. The critical conclusion is that genetic variation in both ovulation rate and embryo survival was present in the base population.

In addition to changes in the number of eggs shed at natural oestrus, genetic selection for traits such as litter size and body weight has often led to changes in the number of eggs shed in response to the injection of pregnant mare's serum gonadotrophin (formerly abbreviated to PMSG, but now preferably to eCG). The extent to which differences in the response to eCG were greater or less than differences in the number of eggs shed naturally has sometimes been used in attempts to assess the extent to which differences in ovulation rate have arisen from genetic variation in the intensity of gonadotrophic stimulation or from genetic variation in the sensitivity of the ovary to that stimulation. The division is, however, somewhat artificial, as the response to eCG must be partly dependent upon endogenous gonadotrophic stimulation. Not surprisingly, therefore, most of the studies indicate a positive relationship, but the genetic correlation is less than one. Some workers have tried to overcome the problem of endogenous gonadotrophic stimulation by hypophysectomy, but even then the response to exogenous gonadotrophins cannot be separated from the stimulation that has taken place up to the time of hypophysectomy. Again, positive correlations have been reported.

Academic interest in the genetic correlation between natural and induced ovulation rate has declined with the realization that the induced ovulation rate was not a measure of ovarian sensitivity alone and with the advent of sensitive assays which have enabled gonadotrophins to be measured directly rather than their effects inferred by difference. Practical interest has, however, increased over the last 10 years or so with the development of intensive programmes of reproduction in domestic animals which rely in part on the use of eCG. The greater the extent to which the correlation is less than one, the greater the extent to which induced ovulation has to be incorporated into an index for the genetic selection of appropriate animals for intensive reproduction. Fortunately, in sheep, those with high natural ovulation rates tend to have high induced ovulation rates, thus indicating a favourable genetic association.

Direct selection

Natural ovulation. The possibility that direct selection for ovulation rate might lead to more rapid changes in litter size than would selection for litter size has often been discussed and advocated as a means of improvement of litter size in domestic animals. For such a procedure to be successful, ovulation rate would have to be a 'better measure' of litter size than litter size itself. Initially such a possibility might seem absurd, but it is genetic merit for litter size we wish to measure, and we already know from the low heritability of litter size that the number of young produced is a poor guide to this. Most of the variation in litter size is determined by the environment. Indeed, so much so that litter size on one occasion is not even a good guide to litter size on another occasion; the repeatability of the trait is low. Discrete variation and small numbers are part of this deficiency. In sheep, for example, if the genetic merit for litter size were 1.5, a ewe would be expected to give birth to singles half of the time and twins the other half. When environmental variation is superimposed – the mean litter size of Scottish Blackface sheep can vary from the minimum of one to over two depending on the level of nutrition – it is not surprising that the number of young at birth is a poor estimate of genetic merit for litter size.

For ovulation rate to be a better estimate of litter size than litter size itself, it must be both the principal source of genetic variation in litter size and more highly heritable. In algebraic terms, everything else being equal, the product of the square roots of the two heritabilities and the genetic correlation must be greater than the heritability of litter size, i.e.

$$h_1 h_2 r_g > h_1^2$$

when h_1^2 and h_2^2 are the heritabilities of litter size and ovulation rate respectively and r_g is the genetic correlation between the live traits.

Direct selection for ovulation rate itself has been studied in two species, mice and pigs, and large responses have been reported in both. In mice, the heritability was found to be 10 and 30 per cent in two different populations. The difference could be partly due to sampling and partly to the fact that Bradford, with the 10 per cent result, made his selection on the number of corpora lutea in the ovaries of females having their first litter, while the study in Falconer's laboratory was based on the number of eggs shed and counted in the ampullae of females in oestrus, following the removal of their first litter. Despite a divergence for over six eggs in both experiments, neither group reported a change in litter size after ten generations of selection. Bradford, however, continued selection for another four generations and then kept the line without selection; during this time the litter size increased gradually and significantly, until after a total of 30 generations the litter size of this line was one-and-a-half young above the control. Presumably, natural selection among embryos had led to the elimination of embryo mortality alleles from the population.

Selection for ovulation rate in the pig was conducted by Dwane Zimmerman and colleagues at Nebraska and based on the number of corpora lutea observed 9–11 days after the second oestrus of a gilt's life. The response was again distinct with over five corpora lutea after ten generations of selection. The pattern of response (Fig. 3.7) illustrates another difficulty when interpreting genetic selection experiments. The ovulation rate of the selected line increased from 14.3 to 16.2 eggs, while that of the control decreased from 14.6 to 13.7, over the first five generations. If the drop in the number of eggs shed by control animals reflected changes in the environment or inbreeding, the 'true response' of the selected line was the deviation between it and the control line. The drop in the control line might, however, have arisen by drift, in which case the 'true response' would be the change in the selected line, i.e. two rather than three eggs.

Again, favourable changes in litter size have not been confirmed, but the observed superiority of the selected line of about half a piglet, although not statistically significant, would be of considerable agricultural relevance if real. Remember also that at this stage Bradford would have concluded from his study that selection for ovulation rate has not increased litter size, yet 20 generations later the improvement in litter size was quite clear. The extrapolation from one population to another, or even from one period

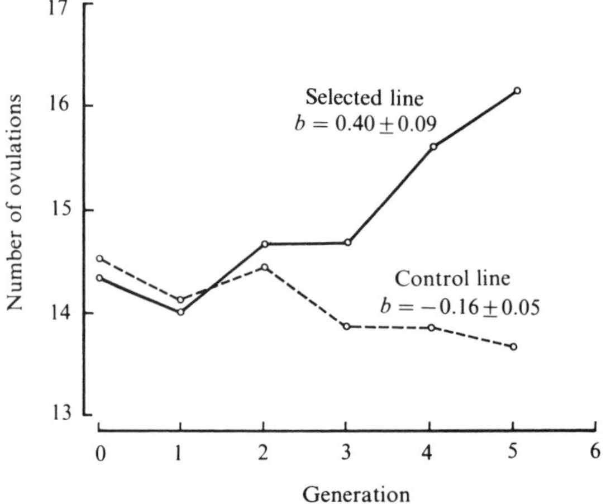

Fig. 3.7. The response to selection for increased ovulation rate in pigs and the regression of the responses on generation number. (After D. R. Zimmerman and P. J. Cunningham. Selection for ovulation rate in swine: population procedures and ovulation response. *J. Anim. Sci.* **40**, 61–9 (1975).)

of time to another in the same population, must be made with caution, especially when considering correlated traits.

These experiments have therefore shown that, in each of the populations chosen, alleles were segregating for the control of ovulation rate. There is no doubt of the existence of genetic variation within these populations and no reason to believe that this variation is peculiar to them. The experiments, however, have not yet demonstrated that ovulation rate might be preferred to litter size as a criterion of genetic selection for litter size. It might be argued that such a result was least likely for the polytocous mouse and pig. For mice the experiments have shown the heritability of ovulation rate to be around twice that of litter size. For pigs the estimate is four or five times that of litter size and there might well be the extra half pig!

Might ovulation rate be a more appropriate criterion for selection in sheep and cattle? Certainly the low ovulation rates are much easier to score and laparoscopy is an appropriate technique. The transfer of additional embryos increases litter size much more convincingly in sheep and cattle than in pigs. A threefold increase in litter size is common in sheep and possible in individual cattle, so that the number born may go beyond the normal range for the breeds studied. Recently Seamus Hanrahan, of Eire, has argued convincingly that all the genetic variation in the litter size of sheep arises from genetic variation in ovulation rate. Environmental variation in embryo mortality therefore acts merely to blur the genetic expression of litter size, which together with the possibility of making repeated measurements of ovulation rate by laparoscopy over a short period of time leads him to conclude that selection for ovulation rate would be two to three times more effective than selection for litter size itself.

Hanrahan himself is selecting to increase the ovulation rate of Finnish Landrace sheep, a breed that already sheds the exceptionally high number of three to four eggs. The trait is responding to selection with a heritability of 0.5, so that even though the population started off at an unusually high value, genetic variation was still present. A strain of sheep with a very high ovulation rate would be enormously valuable to industry for the improvement of other types by crossing. The higher the ovulation rate, the lower the proportion of genes that has to be introduced, and hence the smaller the change in the favourable characteristics of the otherwise desirable breed.

It is a sad reflection on the organization of science and its application that no one is selecting for ovulation rate in cattle or in the principal breeds of sheep. Research has shown that it can be changed in mice, pigs and sheep with very high initial levels. Research has shown that the greatest rewards would be expected in cattle and ordinary sheep. Development agencies have yet to recognize this.

Induced ovulation. The number of eggs shed in response to eCG has also been the subject of direct selection and shown to respond strongly. By comparing the correlated changes in natural ovulation in response to induced ovulation and vice versa we should, in theory, get a better indication of the extent to which variation in the two measures of ovulation rate are determined by segregation at the same loci than we can obtain from the comparison of strains and breeds of unknown history. Unfortunately, the results of the two studies are quite different. In Bradford's experiment there was little evidence of change in the number of eggs shed at induced ovulation, whereas in Edinburgh selection for both natural and induced ovulation rate led to direct and reciprocal changes. The differences in ovulation rate are given in Table 3.3, together with the estimated genetic correlation.

Again, we have the problem of interpretation; one experiment indicated a positive correlation, the other no correlation at all. Current knowledge of the effects of genetic drift, however, argues against the idea of a specific, literal interpretation. We now realize that there may well be a positive correlation, but the two traits are to some extent at least, and possibly to a large extent, under independent genetic control and this type of answer is the most we can reasonably expect. The implications for the genetic improvement of reproductive performance under conditions of exogenous hormonal control are considerable. Animals chosen for their high ovulation rate at natural mating may, but cannot be presumed to, perform well in response to eCG. Knowledge of the genetic control of the components of reproductive performance, therefore, not only helps us to understand the biological control of the trait but may also guide commercial decisions.

Embryo survival. The correlated changes in response to selection for other traits showed that changes in gene frequencies could lead to a decline in embryo survival, and we shall see later that this knowledge is supported by the reduction of embryo survival on inbreeding. Bradford, however, has shown directly that it is possible to increase embryo survival by genetic selection. He used the ratio of implantation sites to corpora lutea in mouse sibs to select for survival, and weighted the ratio by the number of corpora lutea to prevent selection for low ovulation rates. Over 16 generations of

Table 3.3. *The natural and the induced ovulation rates of lines of mice selected for one or the other trait. The arrows point from the direct to the correlated response*

	Natural ovulation		Induced ovulation
Selection for:			
Natural ovulation	7.00	→	4.50
Induced ovulation	2.25	←	16.00
Genetic correlation		0.33	

selection, the proportion of embryos surviving increased from 0.8 to 0.9 and has remained at such a high level through 30 generations of random mating. The number of eggs shed also increased somewhat in this line, so that after ten generations litter size had increased from 8.5 to 10.5.

Selection during those 16 generations has therefore presumably eliminated genes with deleterious effects on prenatal survival from the population. The decrease in embryo loss to an average of only 10 per cent has led to survival of all embryos in 50 per cent or so of the litters in this line. Bradford transferred embryos between the high embryo survival line and the controls, and found that the high survival rate of the selected line was principally a characteristic of the mother rather than of the embryo. The demonstration that survival can be increased through changes in the mother rather than changes in the embryo has enormous implications for our understanding of the causes of embryo loss.

One hypothesis described in more detail in Book 2, Chapter 5 (in the Second Edition) proposes that embryo loss represents the removal of inevitable genetic defects. The development of all ova to young at birth in such a high proportion of litters shows, at the very least, that such genetic damage is not inevitable and casts serious doubts on the generality of this hypothesis. This experiment illustrates the way in which genetic studies can increase our understanding of physiology; the demonstration that embryo loss is not inevitable shows that we have to seek a physiological explanation of the mortality that takes place, and cannot simply hide behind the hypothesis that it is the elimination of genetically defective embryos. We shall return to the interaction between genetics and physiology in the advancement of our understanding of biology later in this chapter.

Inbreeding and crossing

Litter size, as we saw earlier, is depressed by inbreeding; the examination of mice in particular with different degrees of inbreeding of the mother and of the fetuses has enabled the effects of inbreeding to be partitioned into effects on ovulation rate and embryo mortality, and also indicates the nature of the action of the genes affecting the two components.

In general, inbreeding has a greater effect on embryo survival that on ovulation rate, and on pre-implantation mortality than on post-implantation losses. For example, Bradford found that the mean ovulation rate, pre-implantation survival and post-implantation survival of three lines of mice inbred from F_1s were 99, 66 and 86 per cent, respectively, relative to their parents (see Suggested further reading). It is much more difficult to make a simple generalization about the relative importance of inbreeding of the mother and of that of the litter. Inbreeding of the dam probably has the greater effect, but both have been reported. In one experiment there was an interaction between the two; inbreeding of the litter had a much greater effect in outbred than in inbred dams.

It has been concluded from the lack of effects of inbreeding on ovulation rate of mice that most of the genetic variation is additive, that is the effects of a gene are not influenced by other genes, and this conclusion is supported by the fact that crosses among lines tend to be intermediate between their parents. In contrast, the depression of survival-to-birth with inbreeding indicates interactions, and the similarity of crosses to the better parent suggests dominance of the good-survival alleles over those for death. These comments are, however, generalizations and it is important to remember that there are exceptions. In pigs, for example, ovulation rate has been found to decline on inbreeding in some experiments, and the low ovulation rate alleles have been found to be recessive.

Although the nature of the genetic control of the components of litter size cannot be presumed for an uninvestigated population, this does not affect the generalization that litter size shows heterosis.

Genotype × environment interactions

The genes an animal inherits determine not only whether a trait is affected by the environment, and the proportion of overall variation that arises from environmental effects, but also the way in which and the extent to which a trait responds to the environment. When genetically distinct groups respond to a given environmental effect in different ways, the genotype and the environment are said to interact, and such interactions are well documented for ovulation rate. The effect of feeding to a given level of body condition or fatness is illustrated in Fig. 3.8. The South

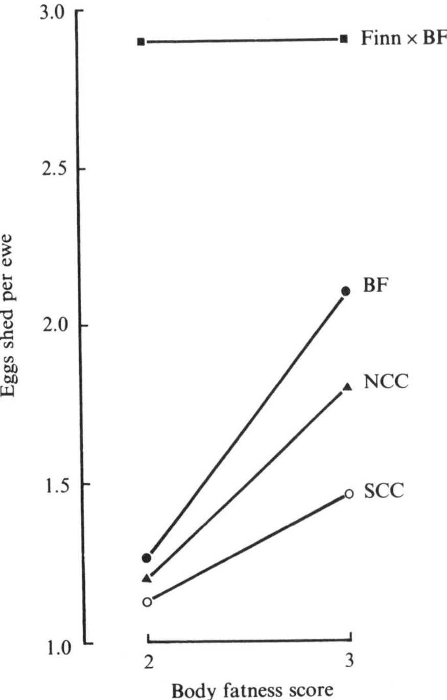

Fig. 3.8. The change in number of eggs shed by Blackface (BF), Finnish Landrace-cross-Blackface (Finn × BF), North (NCC) and South (SCC) Country Cheviot sheep with increasing levels of body fat. (After J. M. Doney and R. G. Gunn. Progress in studies on the reproductive performance of hill sheep. *Hill Farming Research Organization, Sixth Report*, 1971–3, pp. 69–73.)

County Cheviot and Finn × Blackface sheep change their ovulation rate little in response to this change in environment, whereas the Scottish Blackface sheep increase their ovulation rate from 1.25 to 2.1. Sheep of this breed are commonly kept under severe conditions, and an ability to control pregnancy and match the number of lambs born to the food supply could be seen to be an advantage.

Physiology of variation

In the same way that the number of eggs shed has appealed to the geneticist as a character for study as a discrete component of litter size, the number of oocytes present in the ovary has appealed for study as a component of ovulation rate. The aim has often been to try to understand the basis of variation in ovulation rate in terms of variation in ovarian anatomy and variation in the endocrine control of the ovary. The interrelations between the trophic hormones and the maturation of ovarian follicles makes the division somewhat arbitrary, but it nevertheless serves as a useful focus of attention and in addition enables us to describe those interrelations in specific terms.

Oocyte populations

Not surprisingly, artificial selection has not been based on the number of oocytes present in the ovary, and genetic studies have taken the form of the comparison of genetically distinct breeds or selected lines within species.

Contrary to all expectations, even when the total number of primary oocytes has been compared with the number of eggs shed, the relationship

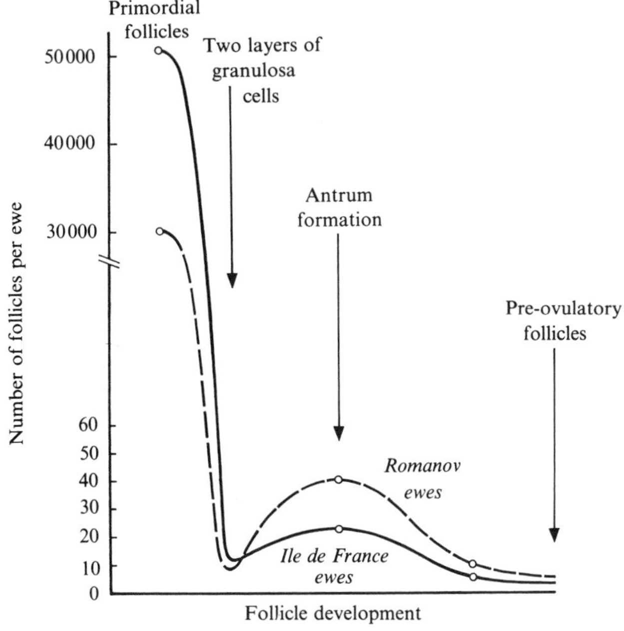

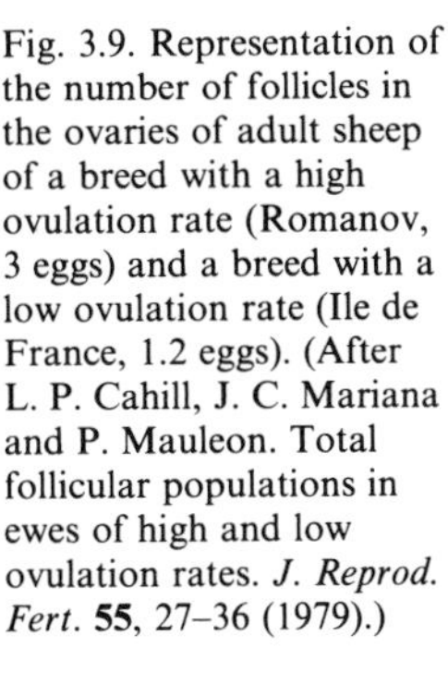

Fig. 3.9. Representation of the number of follicles in the ovaries of adult sheep of a breed with a high ovulation rate (Romanov, 3 eggs) and a breed with a low ovulation rate (Ile de France, 1.2 eggs). (After L. P. Cahill, J. C. Mariana and P. Mauleon. Total follicular populations in ewes of high and low ovulation rates. *J. Reprod. Fert.* **55**, 27–36 (1979).)

has been found to be negative. Strains of mice, rats and sheep with high ovulation rates have been found to have a smaller number of oocytes both at birth and later in life. The number of larger follicles is, however, positively related to the number of eggs shed. The results of one such study are given in Fig. 3.9, as an example. Several people have tried to explain this in terms of follicle dynamics, but as yet no consensus has emerged. In descriptive terms, recruitment from the stock of primordial follicles is greater, and the rate of passage through the final stages of maturation is possibly slower, in animals with high ovulation rates. There is, however, no indication of reduced atresia. The intriguing question of how a lower number of primordial follicles might be related to higher ovulation rates therefore remains. Are the numbers correspondingly lower throughout fetal life or does the lower number represent greater prenatal losses? Might the endocrinology of high fetal loss be the endocrinology of high adult ovulation rate? One thing is certain, however: these relationships represent averages and they represent tendencies. Although the average number of primordial follicles in the Romanov ewes was half that in Ile de France ewes, the range in both breeds was enormous. Romanov ewes varied from 12000 to 58000, Ile de France ewes from 20000 to 86000. *On average* Romanov ewes with a total of 30000 follicles shed more than twice as many eggs as Ile de France ewes with a total of 56000; yet *individual* Romanov ewes still shed twice as many eggs as *individual* Ile de France ewes with 20000! Exactly the same overlap is observed in lambs: some individuals of an inheritance that destines them to have a high ovulation rate have more follicles than lambs of a low ovulation rate inheritance, yet *on average* the relationship is the other way round.

In summary, then, the remarkable consistency among species and studies indicates that there really is a general tendency for high ovulation rates to be associated with smaller numbers of follicles. Given that the same number of eggs can be shed against a fourfold variation in total follicle numbers, the fine genetic tuning of ovulation rate must be mediated through the physiological control of the final stage of follicle development. This we shall now consider.

Endocrine variation

Sensitive methods of hormone assay developed in the late 1960s and applied to the study of genetic variation in the 1970s have transformed our understanding of the physiology of variation in reproductive performance. This was led by Jean Pelletier with LH in 1968. Much of the genetic investigation has been conducted in sheep, a species that has offered a particularly appropriate combination of variation and ease of sampling.

Ovulation rate. Ovarian anatomy is one part of the story, but what causes a larger proportion of oocytes to develop to maturity and ovulate in strains and species with high ovulation rates? Initially it might seem surprising that species differences have not been explored, but the difficulty of extrapolating from between to within species, and the possibility of non-causative differences in particular, may be part of the reason. Humans and sheep, for example, tend to shed one egg at ovulation, yet the concentration of oestradiol-17β in peripheral plasma is of the order of 100 times greater in women than in sheep. This might suggest that genetic variation in ovulation rate was independent of genetic variation in oestrogen metabolism, yet we shall see later that covariation between the two might be the key to the control of variation in fecundity in the sheep.

Sheep studies have centred around the laboratory of the Institute Nationale de Recherche Agronomique at Nouzilly in France, where genetic variation in Romanov sheep, with their unusually high ovulation rates, was first correlated with blood levels of luteinizing hormone in 1970. Surprisingly, results have consistently shown that the Romanov ewe develops her three or so eggs in the presence of concentrations of LH that are largely indistinguishable from those that lead ewes of other breeds, such as the Ile de France, to shed only one or two eggs. More recently, the laboratory has included the measurement of FSH in their studies, and it is intriguing to note that the Romanov maintains luteal levels of this gonadotrophin throughout a greater part of the final phase of follicular growth and development than does the Ile de France (Fig. 3.10*a*).

Oestrogen is known to influence the release of gonadotrophins from the pituitary gland, and there is now evidence to suggest that the highly prolific ewes have a greater concentration of oestrogen in peripheral plasma. This comes from two sources: studies of the rate of secretion of oestradiol-17β into the ovarian vein, and the measurement of concentration in peripheral plasma. The latter would seem the obvious approach, but the concentration is so low that it is very difficult to measure. Nevertheless, the French group have estimated the concentration to be 10 pg/ml in the Romanov compared to 7 pg/ml or so in the Ile de France (Fig. 3.10*b*). What is noticeable is the very much higher concentration immediately before and during the LH discharge in a Romanov. The concentrations of FSH and of oestrogen are both higher in the Romanov during the final period of follicular growth.

The presence of higher levels of oestrogen in the more prolific breed is supported by the greater rate of secretion of oestrogen. Finn–Merino ewes with an ovulation rate of 2.0 secreted two to three times as much oestradiol per minute (1 ng) as did Merinos with their ovulation rate of 1.0 (0.4 ng). Perhaps highly prolific sheep shed more eggs because they are more insensitive to feedback inhibition from oestradiol, thereby allowing them to maintain adequate gonadotrophin secretion in the presence of a greater number of developing follicles. Indeed, it has been shown that daily injections of 50 μg oestradiol block ovulation altogether in Blackface sheep, with their ovulation rate of around 1.5, whereas Finnish Landrace sheep simply drop their ovulation rate from 3.0 to 1.5. This brings us to a very important interaction between genetics and physiology. Physiology has helped us to understand genetic variation but the existence of genetic differences has helped us to identify a possible causative component of variation in ovulation rate. In this vein, the greater concentration of prolactin in the breed with the higher ovulation rate may be a first clue to a role of prolactin in the selection and development of the pre-ovulatory follicle.

The use of conventional physiological techniques led to the supposition that ovulation rate may be related to the concentration of gonadotrophins. If you give additional gonadotrophins, animals from mice to cattle shed more eggs; if you eliminate gonadotrophins by hypophysectomy, folliculogenesis ceases. The initial conclusion seemed reasonable and indeed is still acceptable to the extent that gonadotrophins are essential for gametogenesis. It is the quantitative relationship between the two that remains to be established; the feedback side of the control loop may determine the number of follicles that develop and the number of eggs shed.

The possibility that gonadotrophin release in prolific animals may be less sensitive to negative feedback has several implications. It suggests that

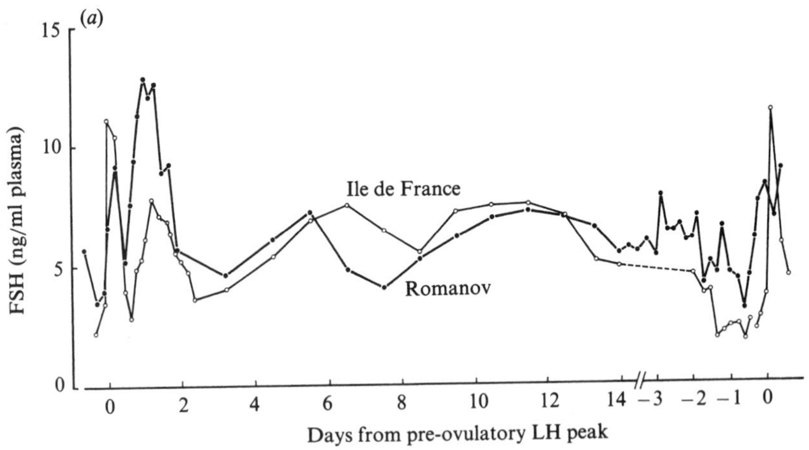

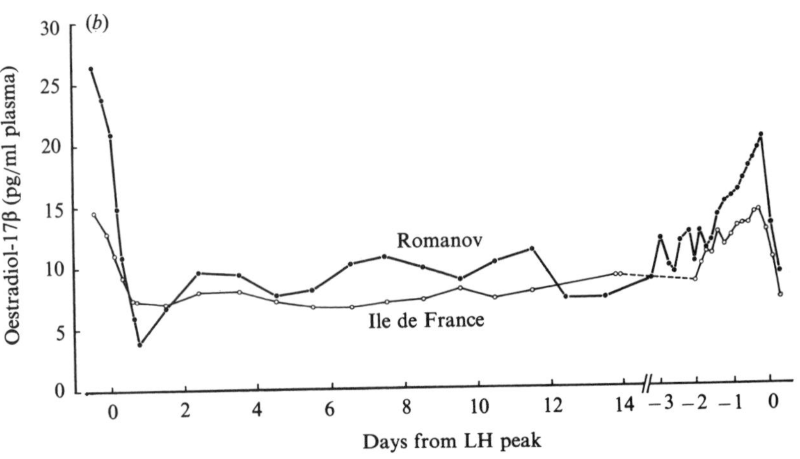

Fig. 3.10. (*a*) The concentration of FSH in the blood throughout the oestrous cycle of high ovulation (Romanov) sheep (solid circles) and low ovulation (Ile de France) sheep (open circles). (*b*) The concentration of oestradiol-17β in the blood throughout the oestrous cycle of high ovulation (Romanov) sheep and low ovulation (Ile de France) sheep. (All data after L. P. Cahill, J. Saumande, J. P. Ravault, M. R. Blanc, J. Thimonier, J. C. Mariana and P. Mauleon. Hormonal and follicular relationships in ewes of high and low ovulation rates. *J. Reprod. Fert.* **62**, 141–50 (1981).)

it might be possible to raise the ovulation rate artificially by interfering with this system; it would mean that genetic variation in the physiological control of ovulation rate might be common to both sexes; it complicates the hypothesis that seasonal anoestrus arises from high sensitivity to negative feedback; and it might enable us to link the tendency for high ovulation rate to be associated with high body growth within a species. Let us consider these implications in turn.

Increases in ovulation rate have now been reported for sheep following immunization against inhibin or gonadal steroids and the use of chemical anti-oestrogens like clomiphene. To refine these techniques and extend them to other species, however, needs further study of genetic variation among species in the feedback effects of different gonadal steroids and of inhibin. The intrinsic merit of this approach is that it would give a buffered change without the increase in variance associated with the use of gonadotrophins such as eCG. Having discovered that natural variation may be compared to different settings of a 'gonadostat', we have the option to try to mimic this by artificial manipulation.

If genetic control of ovulation rate were to be expressed in both sexes it would have an enormous impact on the design of selection programmes and on the rates of response that might be effected. If the degree of genetic merit for this trait could be measured as accurately in males as it can in females, i.e. if the genetic correlation between prolificacy and the endocrine characteristic of the young male were one, the rate of response to genetic selection could be approximately doubled. Fig. 3.11 shows the magnitude of the gains that could be made when the genetic correlation is in the range 0.3 to 0.7. The measurement of the expression in young animals, and

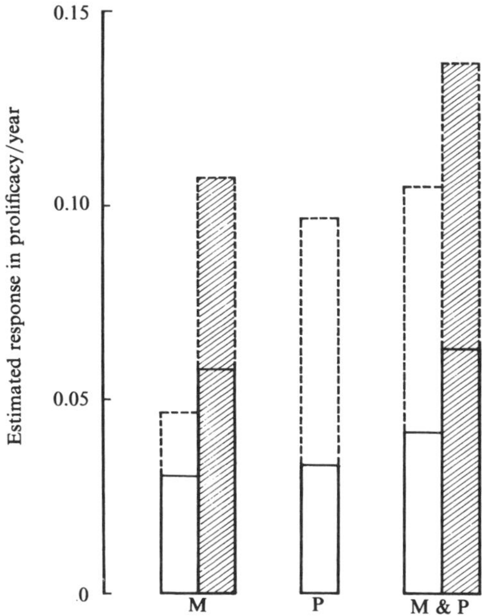

Fig. 3.11. The estimated rate of change in prolificacy when selected alone (P), in response to selection for a male trait alone (M) or in response for prolificacy itself and the male trait (M & P). Solid lines correspond to a heritability of prolificacy of 0.1, broken lines to a heritability of 0.3. The heritability of the male trait is presumed to be 0.3 and the genetic correlation between it and prolificacy to be 0.3 (open column) or 0.7 (hatched column). (After J. R. W. Walkley and C. Smith. The use of physiological traits in genetic selection for litter size in sheep. *J. Reprod. Fert.* **59**, 83–8 (1980).)

especially the expression in the opposite sex, of alleles that normally control variation in the reproductive function of adult females would therefore not only be intriguing in its own right, but also of direct relevance to animal breeding. We shall review the evidence for male–female covariation in the next section.

The use of sensitivity to negative feedback to explain anoestrus and variation in ovulation rate draws attention to the fact that it is really a descriptive operational characteristic rather than a physiological process. Sensitivity could appear to vary because of changes in intrinsic hypothalamic activity or because of changes in the ability to perceive feedback signals. Knowledge of the physiology of genetic variation in seasonality is, however, very sparse, and further studies of both types of variation are required before these can be differentiated. The immunization against oestrogen, which increased the ovulation rate of cyclic sheep, does not, however, affect the incidence of ovulation in response to GnRH during anoestrus. Although we might use the same description for the basis of variation in seasonality and in ovulation rate, the sheep may control the two components of reproduction quite independently; new techniques such as immunization may start to help us to understand this separation.

Advances in the physiological understanding of growth have paralleled advances in the understanding of variation in reproduction. The release of growth hormone from the pituitary gland is normally inhibited by a hypothalamic hormone known as somatostatin, immunization against which may increase growth rate. Hypothalamic failure to signal somatostatin release in response to growth and to inhibit GnRH release in response to folliculogenesis could therefore be the basis of the genetic correlation between body size and ovulation rate within species (Fig. 3.12). Across-species natural selection has eliminated this covariation and established equilibria within species.

Survival of embryos. Since there is little genetic variation in embryo survival, physiological studies have been very limited. One of the few has been the study of Bradford's mouse strains, and the results are summarized

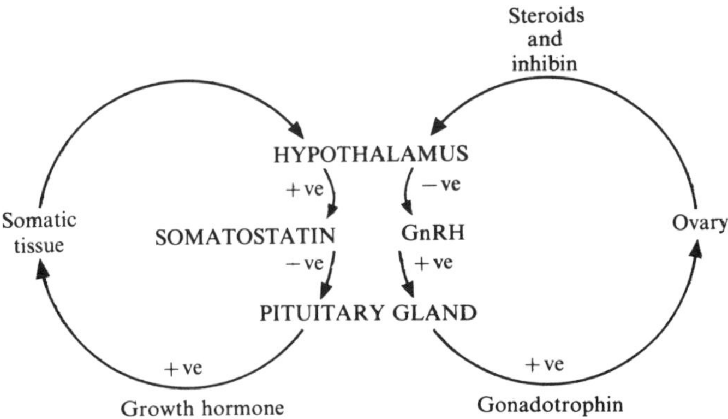

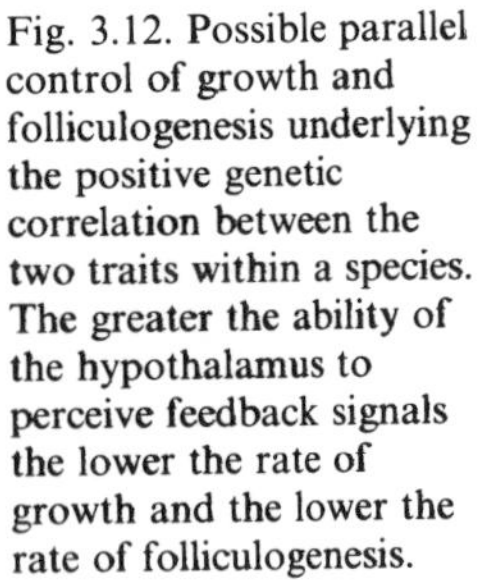
Fig. 3.12. Possible parallel control of growth and folliculogenesis underlying the positive genetic correlation between the two traits within a species. The greater the ability of the hypothalamus to perceive feedback signals the lower the rate of growth and the lower the rate of folliculogenesis.

in Fig. 3.13. Strains with high survival had greater peripheral concentrations of both gonadotrophic hormones and ovarian steroids than strains with low survival. Again, genetic variation may help us to understand causative relationships; the high embryo loss of sheep actively immunized against ovarian steroids might arise in part from the loss of essential steroid action after ovulation.

Male reproduction and male–female covariation

Many fewer males than females are usually required for the reproduction of polygynous mammals and genetic variation in reproductive performance of males has received correspondingly less attention. Several of the reproductive characteristics of males are, however, undoubtedly heritable and this is well illustrated by the enormous variation in reproductive characteristics among males of different species. Specific chapters are devoted to male reproduction (in Books 1 and 3 of the Second Edition) and so here we shall consider the evidence for variation within species. But we shall also pursue the question of the extent to which male reproductive characteristics have evolved under natural selection to meet the requirements of females (as discussed for primates in Book 8, Chapter 1) and, in particular, follow up the possibility that some genes may influence variation in reproductive characteristics in both males and females.

Variation within species

Some of the clearest examples of variation in male fertility come from the effects of single genes. One is the *t* series of alleles in mice, where individuals carrying a mutant *t* allele have distorted segregation ratios and may be sterile. An interesting characteristic of *t* allele sterility is that it is the genotype of the spermatozoon that is affecting that spermatozoon. It is

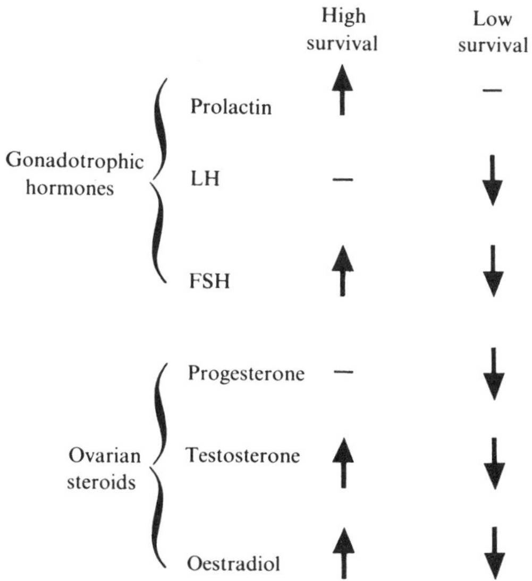

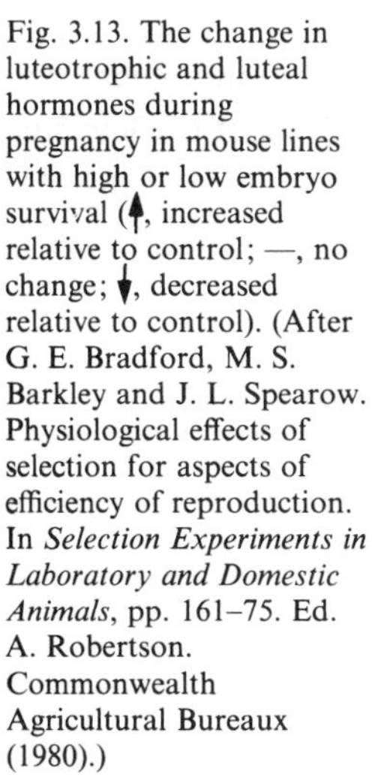
Fig. 3.13. The change in luteotrophic and luteal hormones during pregnancy in mouse lines with high or low embryo survival (↑, increased relative to control; —, no change; ↓, decreased relative to control). (After G. E. Bradford, M. S. Barkley and J. L. Spearow. Physiological effects of selection for aspects of efficiency of reproduction. In *Selection Experiments in Laboratory and Domestic Animals*, pp. 161–75. Ed. A. Robertson. Commonwealth Agricultural Bureaux (1980).)

one of the very few cases of haploid gene expression, for in general the phenotype of the spermatozoon is determined by the genotype of the testis. It is perhaps not surprising that people have so far been unsuccessful in their attempts to separate Y- from X-bearing spermatozoa on the basis of the physical characteristics of the spermatozoon when there is so little evidence to suggest that the genes carried influence the phenotype (see Book 2, Chapter 3, in the Second Edition).

Other examples of the effects of single genes on spermatozoa and male fertility are to be found in cattle, where the widespread use of artificial insemination has led to the careful study of both fertility and sperm morphology. In Friesian cattle, for example, the presence of 'knobbed sperm' arises from an autosomal recessive gene and causes sterility. Similarly 'detached sperm heads' in Guernsey cattle and 'abnormal acrosomes' again in Friesians are inherited abnormalities associated with reduced fertility.

Sperm characteristics in the mouse have also been found to be under additive genetic control. Indeed, 75 per cent of the total variation in several of the dimensions of spermatozoa is controlled by segregation of alleles with additive effects. When the length of mid-piece was changed by divergent selection, discrete lines were formed where the difference in length (23 μm versus 22 μm) was accounted for by different numbers of gyres of mitochondria (three to four). Despite use of the powerful technique of mixed insemination, where spermatozoa 'compete for the ovum' after the female is inseminated artificially with equal numbers of spermatozoa from both types of male, the fertilizing power of the two types of spermatozoon could not be distinguished. We shall refer again to the power of genetic experiments to test the significance of physiological and anatomical characteristics. Clearly, over this range the length of the mid-piece and the number of gyres, and hence size of mitochondria, were of no consequence to the fertilizing capacity of the spermatozoon.

The sperm mid-piece selection experiment also illustrates the influence of both the magnitude of phenotypic variation and the heritability of the trait on the response to genetic selection. Here it was over 50 per cent, the response to selection was very gradual from generation to generation, again indicating little influence of the environment, yet the overall variation in the population was so low that eleven generations of selection in the two lines led to a response of less than 5 per cent of the mean or about 0.2 per cent of the mean per generation. A high heritability alone is not sufficient to ensure a rapid response to genetic selection.

As distinct from the characteristics of the spermatozoa themselves, the number produced is partly influenced by the size of the testes and this in turn is under genetic control. In cattle, for example, the heritability of testis size has been estimated to be 0.67, and the phenotypic correlation between testis size and the rate of sperm production 0.8. Single genes may also affect testis size, or rather cause small testes. Again, in cattle, a recessive autosomal gene has been shown to reduce testis size and sperm production. The gene also reduces the size of the ovary of females and the syndrome is termed gonadal hypoplasia. We shall return to the common genetic control of the gonads of both sexes later in this chapter.

In mice, there are considerable differences in testis weight between inbred strains. At 16 weeks of age, for example, the testes of mice of strain *DBA/2J* weighed 940 mg per gramme of body weight, whereas in the *10J* substrain of *C57BL*, the relative weight was only 390 mg. In one strain (*CBA*), testis size was found to be reduced by a gene on the Y chromosome but it seems to be specific to that strain. As in cattle, small testes reduce fertility.

Direct artificial selection for fertility itself has not been studied in males, so the evidence for genetic variation arises principally from its association with other characteristics such as sperm defects and testicular hypoplasia. As such it tends to represent 'abnormal variation'. The presence of genetic deviants giving a phenotype of subnormal fitness cannot be used to argue that corresponding supranormal types might be expected to exist. The heritability and extent of normal variation in the fertility of males is a much better guide to the possibility of increasing it. This has been estimated from the conception rates of cattle in New Zealand to be 55 per cent. Males therefore carry genes that affect variation in their own fertility as well as that of their daughters.

In addition to the additive genetic variation in testis size, and hence by implication sperm production, the study of crossbred boars demonstrates the presence of non-additive genetic variation and illustrates yet another caution to be noted when interpreting genetic data. Crossbred boars have been found to have 25 per cent larger testes and 35 per cent more spermatozoa per testis than pure-bred boars. They were also heavier and, when the traits were corrected for body weight, the estimates of heterosis fell to 8 and 14 per cent, respectively, and were no longer statistically significant. Operationally, the crossbred boars did produce more spermatozoa and show heterosis, but whether this reflects non-additive genetic variation for sperm production as such or ability to thrive is not known. To the pig breeder, boars may be thought of as an essential overhead; it is ultimately sperm production per unit of cost that is important, so if costs don't rise linearly with body weight, it is sperm production *per se* that counts and crossbred boars show heterosis.

We shall discuss later the genetic variation in male seasonality in relation to the seasonality of females.

The requirements of females

Species difference in the reproductive characteristics of males may be exhibited naturally, as discussed in other chapters; they may also be brought out by alterations to the environment. Regardless of the relative efficiencies of males of the principal domestic species at individual mating, the use of artificial insemination leads to a vast superiority of the bull over boars, rams or stallions in the number of females inseminated per unit of time. This is illustrated in Fig. 3.14. It is not only that existing technology enables us to dilute and freeze bull semen particularly well, but we are also better able to overcome the barrier to sperm transport in the female. Although the cervix is the principal barrier in both sheep and cattle, the cervix of the cow is less tortuous, so that an insemination catheter can pass to the uterine lumen. In the ewe this is not possible. In the sow, the principal barrier is the utero-tubal junction which cannot be passed by a conventional vaginal approach. Genetic variation in the anatomy of the female and in the site of the barrier to sperm transport therefore determine much of the genetic variation in the suitability of males of these species as sources of semen for AI.

Other physiological characteristics of females may also be important. In rodents, for example, copulation is characterized by a very large number of pre-ejaculatory thrusts. These may number several hundred and mating may be prolonged for 20 min. The evolutionary advantage of such a pattern of behaviour is difficult to recognize until one remembers that mating in the rat and mouse stimulates the formation of a functional corpus luteum independently of conception. But why should such a prolonged stimulus be required? Studies of the effects of the components

Fig. 3.14. The variation among domestic animals in semen characteristics and number of inseminations possible per ejaculate.

Species	Volume of semen obtainable from single service	Sperm per ejaculate $\times 10^9$	Easily frozen	Number of inseminations from a single ejaculate
	50–100 ml 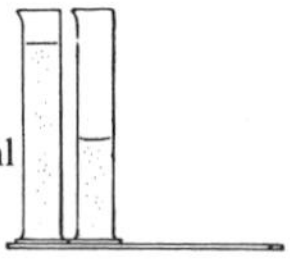	3–15	No	90
	5 ml 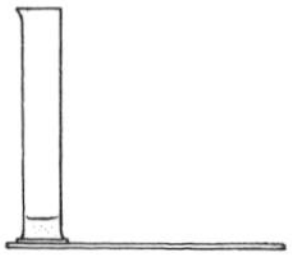	4–14	Yes	500
	1 ml 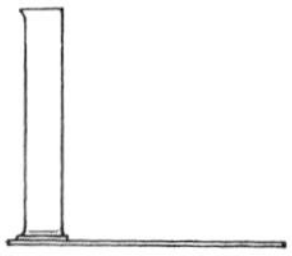	2–4	No	25
	200–250 ml 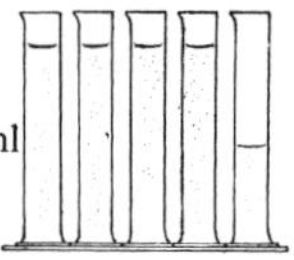	40–50	No	35

of mating show a difference between rats and mice. In the former, four or more of the eleven or so intromissions are required for the induction of a luteal phase in the oestrous cycle. In the mouse, however, it is the final ejaculatory reflex that gives the signal for luteinization, and the pre-ejaculatory thrusts do not seem to have a role. Despite superficially similar patterns of mating in two species of the family Muridae, the functional significance of the components is quite different.

Within species there is again evidence of genetic variation in behaviour. Males of the inbred strain of mice *C57BL* may thrust 400 times before ejaculating, whereas males of the *DBA* strain ejaculate after an average of 125 thrusts! Similarly, in sheep, the libido of some breeds is greater than others and may show differing seasonal variation. Whether this is to accommodate the requirements of the female or whether it is the direct result of similar genes causing variation in both sexes is not known.

Before leaving species differences in male–female interactions, it must be remembered that the spontaneity of the luteal phase of the oestrous cycle of domestic species does not mean that males do not influence female reproduction. Indeed, there is considerable evidence to the contrary. In sheep, mating affects the timing of the LH discharge, and the presence of males may induce oestrus before the spontaneous onset of the breeding season.

Male–female covariation

The breeding season of sheep and the libido of rams is an appropriate point to introduce the hypothesis that variation in the physiology underlying reproduction may be common to both sexes. Some of the first evidence to lead to this hypothesis was the observation that males of breeds where the females had high ovulation rates had greater libido than those of breeds with normal ovulation rates. Furthermore, when the female had an extended breeding season, the male had a corresponding extended period of high libido. It has since been reported that the concentration of testosterone may start to rise in the peripheral plasma of male sheep in North Africa as soon as the days start to lengthen in February. The male is therefore showing a long-day rather than a short-day tendency in the same way that female sheep in this region have been reported to start regular oestrous cycles while the duration of daylength was still increasing. The same caution must be given to genetic interpretations as was discussed earlier for the female, and it may well be that genes that cause short-day breeding at high latitudes cause long-day breeding at intermediate latitudes. Even so, the similar response of males and females indicates that autosomal genes may determine the time of the start of breeding activity – regardless of sex. Maybe we should expect this rather than draw attention to it. After all, both male and female voles tend to breed in response to long days, whereas sheep tend to breed in response to short days. The species differences are compatible with common genes in both sexes, and it is generally accepted that complicated hypotheses should not be considered if simple ones are adequate. It is much simpler to think of natural selection affecting the frequencies of the same genes in both sexes than to consider selection for females to give birth at the most appropriate time, and then for males to be sexually active a gestation period later. After all, there is no *need* for males to have a period of sexual inactivity – it is the females that have to avoid giving birth when the environment is unsuitable.

A common control of the breeding season in both sexes might then reasonably be expected. Similarly, parallel puberty in both sexes is not surprising. Again, in sheep, male lambs of a breed where the females show puberty in their first breeding season grow their testes in that season, whereas when the females delay puberty to the second season of their lives, the males delay the growth of their testes (Table 3.4). The inheritance of puberty is then apparently controlled by genes that show complete dominance, the Finnish Landrace being homozygous for the dominant alleles, the Tasmanian Merino for the recessive alleles. Remember, though, in our discussion of heterosis, how the imposition of an artificial threshold or of a particular environment may transform additivity to apparent heterosis. So too could the imposition of the part-natural/part-artificial limit to the period of mating at 5–6 months of age transform additivity to dominance. Suppose Finnish Landrace lambs mature at 4 months, Merinos at 7 and crosses at 5.5. The narrow 'window' of measurement at 5–6 months converts this additivity to dominance. We can only talk about the mode of inheritance for a particular environment.

The similar genetic control of the breeding season in both sexes and the known common physiology of the stimulation of testicular and ovarian function led to the question of whether the genes that control ovulation rate and hence litter size in a female might be expressed in a male. Many workers have since shown that the concentrations of LH in the peripheral plasma of male lambs of breeds with high ovulation rates are greater than in lambs of breeds of low ovulation rates (e.g. Fig. 3.15). The possibility that alleles that influence ovulation rate in a female environment may be expressed and recognized in a male is now accepted. The problem of how best to recognize them remains.

Table 3.4. *The incidence of oestrus in female lambs at 5–6 months of age and the size of the testes of male lambs relative to their size at 18 months of age*

Breed	Cyclic females (%)	Relative testis size in males (%)
Finnish Landrace	97	85
Tasmanian Merino	5	40
F_1	94	75

One possibility is that the end-product of gonadotrophic activity in the male – testis size – may integrate such activity and reflect alleles for variation in ovulation rate. Selection for testis size in mice has led to changes in ovulation rate, the genetic correlation between the two traits being 0.5. Similarly, in some cases where testis size has been measured following selection for ovulation rate or litter size in mice, it has been found to change in the same direction as the female trait, but this is not always the case. A particular difficulty in applying such knowledge to farm animals is the interaction of testis growth and function with body growth. Larger animals tend to have larger testes, and testicular steroids influence growth. Variation in the genetic relationships between physical maturation and that of the reproductive system are now being investigated within species.

Another possible integrator is the LH response to GnRH. Although the characteristic seems highly heritable, its association with ovulation rate is not established. In one experiment in sheep, high and low selected lines differed by a factor of 2 after only one generation of genetic selection. The trait has both a high phenotypic standard deviation (50 per cent of the mean) and a high heritability (33 per cent).

Such an experiment further illustrates the use of genetics to establish and investigate the physiological 'relevance' of a particular characteristic. If drastic changes in the LH response of the lamb to GnRH leave reproduction unaltered, it may be regarded as irrelevant. Genetic selection can bring about gradual, natural changes in the physiological characteristics, and so in many respects offers a method of investigation superior to the more traditional use of supplementation and ablation.

To complete the link back to females, males with high rates of testis growth are apparently less sensitive to the negative feedback effects of gonadal steroids on gonadotrophic release. The immunization of males

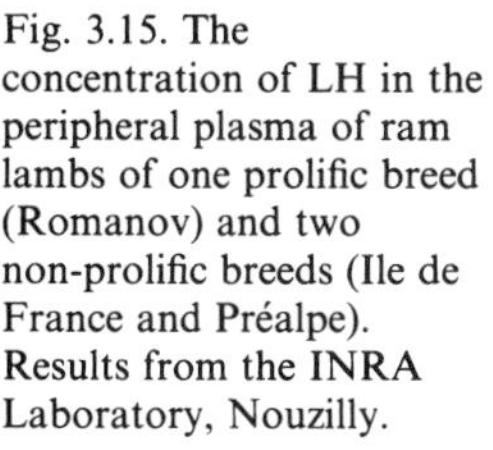

Fig. 3.15. The concentration of LH in the peripheral plasma of ram lambs of one prolific breed (Romanov) and two non-prolific breeds (Ile de France and Préalpe). Results from the INRA Laboratory, Nouzilly.

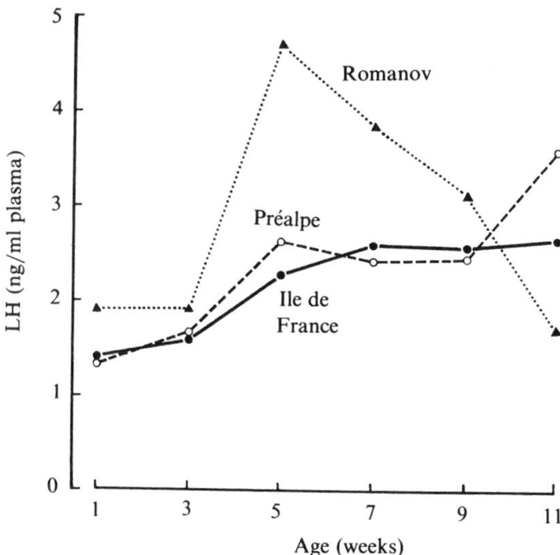

against oestrogens increases the rate of growth of their testes in the same way that it increases the number of eggs shed. The convenient measurement of this characteristic in young males remains a challenge.

While discussing testis size in the context of the requirements of females it has recently been pointed out that among primates there is a positive relationship between testis size and the number of males that mate with each oestrous female. The implication here is that, where natural selection has favoured the acceptance of more than one male by the female, it has also favoured an appropriate testis size to produce an adequate ejaculate. It is therefore possible that testis size relative to body weight may indicate the mating behaviour of a primate species.

A similar correlation has now been found within other species but we must be careful when reversing arguments. Remember that the relationship between body weight and litter size is quite different within and between species. Remember also that the heritability of testis size in sheep, mice and cattle, where it has been studied, is close to one-half. Such a high heritability is quite atypical of a trait subject to natural selection. We need to know the proportion of variation in testis size that is associated with variation in the pattern of mating behaviour and the proportion that is associated with variation in female performance. These are genetic correlations and, with such things, generalizations are fraught with danger.

Pseudoautosomal genes

The presence of a gene for small testes on the Y chromosome has already been mentioned for the *C57BL* strain of mice. Other genes controlling

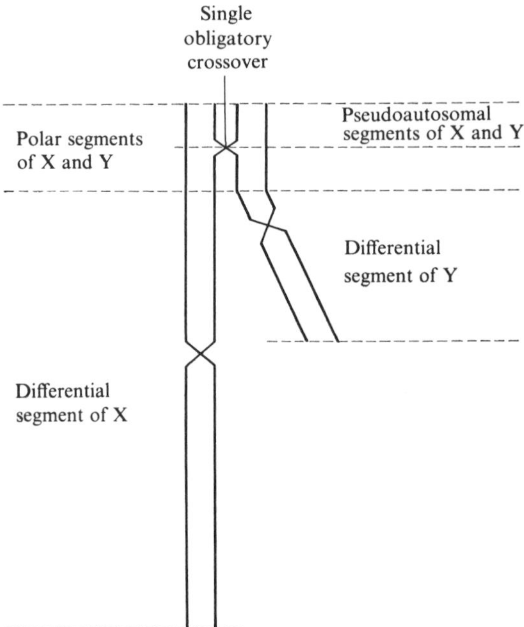

Fig. 3.16. Postulated crossing-over between the terminal segments of the X and Y chromosome at meiosis. Any gene distal to the crossover will therefore alternate between the X and Y chromosome in succeeding generations, and mating studies will fail to detect that it is sex-linked. Such genes are referred to as 'pseudoautosomal'. (After P. S. Burgoyne. *Hum. Gen.* **61**, 85–90, fig. 1 (1982).)

reproduction behave autosomally; genes for female reproduction, for example, are transmitted through the male. The distinction between sex-linked and autosomal genes has, however, been potentially blurred by the hypothesis of Paul Burgoyne that a crossover always takes place between the paired homologous sections of the X and Y chromosome during meiosis in the male. All loci distal to the crossover will therefore behave autosomally, as shown in Fig. 3.16, and he calls such loci pseudoautosomal. Burgoyne has discussed the inheritance of the sex reversal mutation in the mouse (described in Book 2, Chapter 3) and suggests that with no evidence for linkage to an autosome in the mouse, the autosomal nature of its inheritance is misleading; it is located on the sex chromosome but distal to the crossover.

Single genes

The clearest example of the effect of a single gene controlling reproduction is the part of the Y chromosome that determines sex. Further, there are several examples of single genes affecting sexual differentiation, such as sex reversal (*sxr*), testicular feminization (*tfm*) and the polled gene in the goat, and of heteroploidy (unusual chromosome number) having the same effect, such as XO and XXY. (These and the effects of other errors in cell division, are discussed in Chapters 1, 3 and 5 in Book 2 of the Second Edition.) In addition, there are single genes that have distinct effects on the phenotype of the individual without affecting sex. In that it is unusual for single gene substitutions to have such effects – they are not part of the natural distribution of reproductive characteristics – they are by definition abnormal. In that they have such discrete effects they have been ideal for physiological study, and therefore merit discussion as examples of the way in which reproduction might be affected; four of them are to be considered here. Three, the hypogonadal, the obese and the dwarf genes in the mouse, reduce fecundity, while the fourth, the Booroola gene in sheep, increases fecundity.

The hypogonadal mouse (*hpg*)

The hypogonadal mouse was first recognized as a mutant at Harwell but its physiological characteristics have largely been studied in the Department of Anatomy at Oxford. The mutant gene is recessive and homozygotes are sterile, both males and females presenting immature, very small gonads. The intriguing characteristic is that, unlike the pituitary dwarf and the obese mouse, body growth is not affected directly. (Females have similar body weights to normal females, whereas males at 6 weeks are some 20 per cent smaller than normal males, an effect thought to arise indirectly through a reduction in anabolic steroids in the hypogonadal male.) The gene has been found to prevent the production of GnRH by the hypothalamus. GnRH cannot be detected in plasma, the LH and FSH content of the pituitary gland is very low and FSH and LH are virtually undetectable in plasma. Immunologically reactive GnRH has not been found to be present in the hypothalamus.

If GnRH is given, males resume spermatogenesis and females start to secrete LH and FSH. Transplantation of the gonads to normal mice showed that the testes and ovaries are capable of gametogenesis; more recently gametogenesis has been induced in hypogonadal mice by hypothalamic transplants from normal mice. The hypogonadal mouse is then a remarkable example of a specific lesion.

The obese mouse (*ob*)

The obvious characteristic of the obese mouse is its hyperphagic obesity. In addition to its reduced fertility it is also characterized by a reduced ability to generate heat and an inability to maintain body temperature at low environmental temperatures. Although males are characterized by reduced fertility, in the sense that they are fertile for only a brief period after puberty, females are sterile. However, like the hypogonadal mouse, the infertility does not seem to be primarily a gonadal characteristic. Transplantation of ovaries to normal hosts facilitates the completion of folliculogenesis, so that inadequate gonadotrophic stimuli are again implicated. This possibility has been studied at the Department of Anatomy of the Royal Veterinary College, London.

The concentration of LH in plasma seems to be normal, while that of FSH is reduced in both sexes. The pituitary gland releases gonadotrophins in response to the injection of GnRH. The kinetics of GnRH in the hypothalamus, however, appears to be abnormal, for the concentration of GnRH in the hypothalamus is higher than normal in young obese mice but, if anything, lower than normal in old mice. The pituitary gland of obese mice contains less LH than that of normal mice. The data therefore indicate a reduced release of the GnRH from the hypothalamus, and hence reduced stimulation of the pituitary gland. When the reduced activity of the gonad is put together with this depressed release, the lesion in the obese mouse appears to be an excessively high sensitivity of the hypothalamus to the feedback effects of gonadal hormones.

The obese mouse thus contrasts sharply with the hypogonadal mouse. In the obese mouse there is a hypothalamic abnormality which results in the loss of control of metabolic and reproductive function. In the hypogonadal mouse the abnormality is a specific inability to produce GnRH.

The dwarf mouse (*dw*)

The dwarf mouse, like the obese mutant, combines abnormal body growth with abnormal reproduction, but this time homozygous animals are smaller than normal. They appear to grow relatively normally to about 17 days of age, but by 6 weeks of age may weigh only 8 g compared to 25 g for heterozygotes or homozygous normals.

The principal endocrine characteristics are reduced prolactin and growth hormone concentrations compared to normal mice, but FSH and LH also are at a lower concentration in peripheral plasma and are present in smaller amounts in the pituitary gland. The concentration of FSH in plasma at 20 days of age is only 20 per cent of that of normal mice. In many respects the effects seen in the dwarf mutant are comparable to those of hypophysectomy, but antral follicles do develop in dwarf mice and there are some reports of ovulation and even mating. Of particular interest is the greater number of primordial follicles in the ovary – remember that we discussed earlier the negative genetic correlation between ovulation rate and the total number of follicles present in the ovary.

In short, the dwarf mouse may be compared to an exaggeration of a low-ovulation-rate strain: it has more follicles than normal but hardly any of these develop. Like the obese but unlike the hypogonadal mutant it is part of a larger syndrome.

The Booroola Merino

The physiological basis of the effects of the Booroola gene in sheep is less well understood than for the mouse mutants above and it has only recently been recognized as a single gene. In gross terms, it doubles the number of lambs born to Merinos of otherwise low fecundity and, again in contrast to the mutants already discussed, it has a dominant effect on litter size.

The gene was first recognized by breeders in New South Wales who selected on litter size and isolated a strain of high fecundity Merinos. Laurie Piper and Bernie Bindon of the Australian Commonwealth Scientific and Industrial Research Organization have since studied it extensively at Armidale. Lambs have higher LH levels than normal lambs, just like the lambs of the highly fecund Romanov and Finnish Landrace breeds. The gene increases ovulation rate and, interestingly, seems to act additively at this level. Increased loss of embryos in the homozygous Booroola females results in similar litter sizes for homozygotes and heterozygotes and hence the effects of the gene are dominant when measured at this level.

In practical terms the control of the high fecundity of the Booroola Merino by a single gene has advantages and drawbacks. On the positive side it could be incorporated, by backcrossing, into other strains without any serious change in their otherwise desirable commercial characteristics. Unlike a trait with polygenic inheritance, however, it is not possible to take a part of the effect, and an extra lamb might be too much in many extensive husbandry systems.

Drawing conclusions from single genes

It is difficult to strike a balance between the advantages and drawbacks of single genes with large effects for the study of the genetic control of reproduction. Of the mouse genes, one was associated with increased body weight, one with decreased weight and the third with unchanged body weight, yet all were associated with reduced fecundity. We know that in general, however, low fecundity is associated with low body weight within a population. It is therefore important to remember that single genes are examples of what might and can happen, but that it is *not* possible to extrapolate *directly* from them to the nature of the variation and covariation that might be expected in normal populations. The gap between single genes and normal variation is much greater than that between inbred strains and normal variation.

One of the reasons why it has been so difficult to get to the basic action of single genes is that these effects are so large that there are repercussions throughout many aspects of the physiology of the homozygote. According to current theory of recessive gene action, there is a primary deficiency, but the levels of physiological activity 'down gene' are normal. Methods of physiological measurement are now such that it may be possible to detect effects closer to the gene and hence identify the primary action of the abnormal allele. Up to now we have tended to argue that the effects in the homozygote were so large that it 'must' be easy to identify the primary lesion. The time has come to be more subtle. The detection of differences between heterozygotes and homozygous normal animals may present an additional tool in the search to advance our understanding of gene action and the genetic control of the physiology of reproduction.

Genetics is the science of the way individuals reproduce themselves, and in that reproduction is one of the many characteristics of an individual, it is the science of the control of variation in reproduction itself. I hope that this chapter has illustrated the extent of genetic variation in the reproductive patterns of mammals and given an indication of the physiological mediation of that variation. We must not confuse the heritability of a trait with the extent to which that trait as such is determined genetically. The heritability is an estimate of the proportion of observed variation that is caused by additive effects of the genes segregating in that population. The tendency for sheep to produce either one or two lambs at a time is a genetic restriction, yet 80–90 per cent of the variation in numbers born in a population is determined by the environment. Reproductive traits tend to have lower heritabilities than body characteristics such as length or wool type, but the genetic control is stricter. The power of genetic selection is illustrated by the fact that the mean value of selected populations may be outside the range of values seen in the original population.

Knowledge of genetic variation in reproductive performance advances our understanding of why species are what they are, but it also indicates the opportunities open to man to change the reproductive patterns of domestic species to his advantage. We now know it is possible to do this; the question is whether it is worth it. In that genetic change is potentially for ever, it is the simplest form of change, but there must be assurance that change is improvement and that resources for effecting changes are allocated appropriately. To date the emphasis has been on the agricultural improvement of growth characteristics, but in the future the success of these programmes, and the possible adverse effects of changes in performance on reproduction, might both be expected to lead to a greater emphasis on the genetic improvement of the reproductive performance of our farm animals.

The interactions between the physiology of reproduction and the genetic control of reproduction are perhaps the most exciting. The beauty of these interactions is that their benefits are not one-sided. Knowledge of the physiological expression of alleles that have a beneficial effect on reproduction may help us to recognize the presence of such alleles and hence aid genetic improvement. Equally, the availability of genetic differences provides natural variation for physiological study and an ideal way to assess the causal relevance of variation in particular steps of a physiological pathway. The interactions between physiology and genetics, and their different places in the order of control, make a nonsense of the old argument as to whether traits were limited genetically or physiologically. It is a question of genes to code for the appropriate physiology.

The application of the radioimmunoassay developed by Jean Pelletier in 1968 to the study of genetic variation led to a decade where physiology and genetics interacted very effectively to advance our understanding of reproduction. The recent introduction of additional cloned growth hormone genes to the mouse genome heralds a new molecular biology of reproduction, and the prospects of yet another exciting decade of advances in the understanding of the genetics of reproduction.

Suggested further reading

Some notes concerning the genetic possibilities of improving sow fertility. K. Johansson. *Livestock Production Science*, **8**, 431–47 (1981).

Physiological criteria and genetic selection. R. B. Land. *Livestock Production Science*, **8**, 203–13 (1981).

Genetical influences on growth and fertility. R. C. Roberts. *Symposia of the Zoological Society, London*, **47**, 231–54 (1981).

Problems and possibilities for selection for fecundity in multiparous species. O. Vangen. *Pig News and Information*, **2**, 257–63 (1981).

Assessment of new and traditional techniques for selection for reproductive rate. B. M. Bindon and L. R. Piper. In *Sheep Breeding*. Ed. G. J. Tomes, D. E. Robertson and R. J. Lightfoot. Western Australia Institute of Technology; Perth (1976).

Genetic variation in prenatal survival and litter size. G. E. Bradford. XIII Biennial Symposium on Animal Reproduction. *Journal of Animal Science*, **49** (Supplement II), 66–74 (1979).

Selection for increased ovulation rate, litter size and embryo survival. J. P. Hanrahan. *Proceedings of the II World Congress on Genetics Applied to Livestock Production, Madrid.* Editorial Garsia; Madrid (1982).

Genetics of reproduction in the pig. W. G. Hill and A. J. Webb. In *Control of Pig Reproduction.* Ed. D. J. A. Cole and G. R. Foxcroft. Academic Press; Sydney (1982).

Genetics and Reproduction. Journal of Reproduction and Fertility, Supplement 15 (1972).

Further possibilities for manipulating the reproductive process. R. B. Land, I. K. Gauld, G. J. Lee and R. Webb. In *Future Developments in the Genetic Improvement of Animals.* Ed. J. S. F. Barker, K. Hammond and A. E. McClintock. Academic Press; Sydney (1982).

The laboratory mouse as a model for animal breeding: a review of selection for increased body weight and litter size. J. C. McCarthy. *Proceedings of the II World Congress on Genetics Applied to Livestock Production.* Editorial Garsia; Madrid (1982).

Introduction to Quantitative Genetics. D. S. Falconer. Longmans; London (1981).

4

The environment and reproduction

B. K. FOLLETT

Environmental conditions dominate reproductive achievement and animals must be able to adjust to, and even anticipate, the changes taking place around them. This may well require a rapid response such as we see in a male mouse which, entering a new territory, excretes a pheromone that brings the females into oestrus within 3 days, thus allowing him to populate his environment quickly and so pass more of his genes into the next generation. Equally, the reproductive response to environmental change may take years to develop; consider how improved nutrition has slowly but steadily reduced the age of puberty in man, the elephants and the baleen whales.

A more widely occurring event is that of seasonal breeding, an important strategy for maximizing reproductive success and one that by definition is regulated by the environment. The advantages of producing young only at the ideal time are well exemplified by first-year mortality in a colony of rhesus monkeys living on La Parguera, an island off Puerto Rico. Mortality was 22 per cent for monkeys born around the median birth date in June, but rose to 34 per cent if they were born in April and to 52 per cent if born in August. The key factor in determining when young should be produced is normally the availability of food around the time of birth; thus lambs are born in the spring when the growth of new grass can sustain lactation and promote weaning, while birds such as rooks also breed early because the earthworms needed to feed the young become less available later as drier weather forces them deeper into the soil. In contrast, seed-eating finches must hatch their young later when the grasses have ripened, whilst some hawks in Spain delay their breeding until autumn when the birds on which they feed their young arrive overhead, migrating to Africa. The list of specific adaptations is endless but the point is simple: there is an optimal time for giving birth for each species, determined largely by the time of maximum food availability.

Young cannot be produced simply on demand, however, and so preparations must begin well in advance to allow sufficient time for gonadal development, gamete production, the establishment of breeding territories and gestation. Since all this can take anything from 2 months in a small rodent or finch to a year or more in large mammals, it follows that animals are dependent on environmental cues, often called proximate factors, to tell them when to begin the reproductive process. In theory, any environmental factor could be used, but evolution ensures that only the most reliable indicators are selected. Nevertheless, a wide array of environmental factors trigger reproductive competence. For example, giraffe conceive when the acacia trees are in full leaf, and some animals even have built-in annual 'clocks' (see Book 3, Second Edition) which run accurately for months on end and are vital for timing breeding in hibernating ground-squirrels, or in warblers wintering in equatorial rain-forests far from their breeding grounds.

The further one moves away from the tropics, the more pronounced seasonal breeding becomes (Fig. 4.1) and the more precise the timing. Fortunately, the tilt in the earth's axis as it makes its yearly orbit around the sun provides an ideal proximate factor at higher latitudes in the form of the annual changes in photoperiod, and birds and mammals use length of day widely, to regulate not only reproduction but also many other seasonal processes, like hibernation, migration, appetite, hair growth, moulting and colour change. We shall next consider in some detail the proximate factors that control reproduction and some of the many other influences of environment on reproduction.

Photoperiodism

By the early seventeenth century the Dutch already knew that day-length influenced reproduction in birds. In order to capture autumn migrants they required decoys that would sing out of season, at a time of the year when all sexual activity had ceased; they therefore placed birds in total darkness after the breeding season (thus breaking refractoriness, as we shall see below) and then stimulated them by exposure to natural day-lengths. These discoveries had little impact on science, however, and it was only 60 years

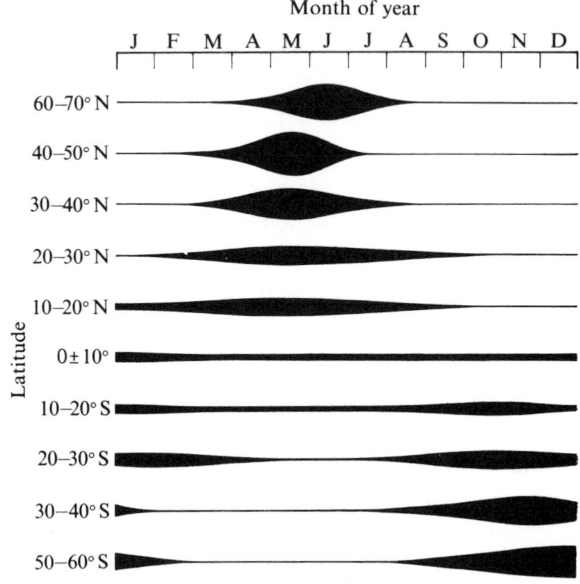

Fig. 4.1. Seasonal breeding in birds nesting at different latitudes. The diagram shows the relative number of times that eggs have been found at each month of the year. (Redrawn from J. R. Baker. *Proc. Zool. Soc. Lond.* **104**, 557–82 (1938).)

ago that William Rowan in Edmonton, Alberta, first analysed the effects of light on gonadal growth in birds. His most detailed experiments were with Oregon juncos: when he treated them for some weeks with extra illumination in the evening, all the birds began to sing and show gonadal growth. Perhaps it was the fact that these birds could be brought into reproductive activity in the depths of a Canadian winter, when temperatures rarely exceed those of a deep-freeze, that convinced sceptics that light, not temperature, was the primary regulator of seasonal breeding. Within a decade, work had been extended to mammals, and the ferret and vole had been shown to be photoperiodic; by now, examples can be quoted from most mammalian orders.

Long-day and short-day breeders

Whether a species reaches full reproductive maturity under long or short days very much depends upon its length of gestation (Fig. 4.2). Most rodents and carnivores, with pregnancies of 20 to 80 days, are long-day

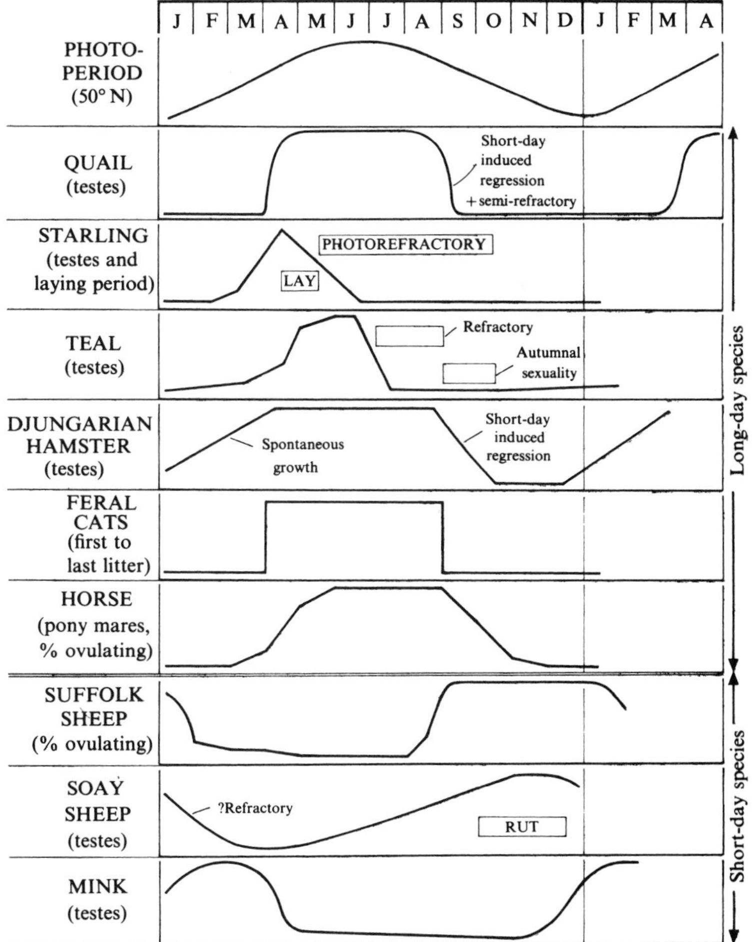

Fig. 4.2. Seasonal gonadal activity in some photoperiodic birds and mammals. (Various sources.)

breeders and mate in the spring and summer. The same is true for virtually all temperate-zone birds. However, within this broad category there are refinements and complexities between species, some of which must involve considerable physiological differences. The cycle of many small rodents, for example, begins in late summer when the shortening days combined with decreasing food supplies and lower temperatures cause the gonads to regress. This photoperiodic inhibition continues for months but then wears off and the gonads grow spontaneously in spite of the prevailing short days. As a result most rodents can breed in the early spring. Long days are stimulatory and maintain the animals in breeding, but they also cause the rodent to regain its sensitivity to the inhibitory effects of short days, so that the following autumn they once again regress.

This explanation has been largely derived from experiments on the golden hamster *Mesocricetus auratus* and the Djungarian hamster *Phodopus sungorus*, but it probably applies to most small rodents, even though details will vary depending upon the individual situation of the species. For example, the degree of photoperiodic control depends upon latitude. Deer mice *Peromyscus maniculatus* and white-footed mice *P. leucopus* exist through much of North America but the sub-species have different reproductive cycles. Year-round breeding occurs in Mexican populations (25–30 °N), reflecting the relatively even annual distribution of food compared with more northerly latitudes, and exposing these males to short photoperiods does not alter testicular size. In contrast, populations from Canada are photoperiodic and the testes regress completely under short days. However, not all Canadian mice come out of breeding in winter and up to a quarter of the population are potentially capable of reproduction should environmental conditions allow.

Domestic cats also illustrate the interaction between reproduction, latitude and long day-lengths. From the equator to 20 °N, feral cats breed at any time of the year, but as the latitude increases they gradually become seasonal, until at 50–60 °N litters are produced only from March until September. This photoperiodic response also occurs in the laboratory. Under a constant 12 h of light and 12 h of darkness (LD 12:12), as at the equator, cats breed continuously with about 4 per cent mating each week; short day-lengths (LD 9:15) reduce the rate to 1 per cent per week, whereas under LD 14:10 it rises to as high as 10 per cent.

Birds have highly evolved long-day responses because their breeding is usually dependent upon the availability of a very specific food source. Typically, in a species such as the starling *Sturnus vulgaris*, the gonads are undeveloped during the winter (testes, 5 mg), but grow once the critical day-length is exceeded in spring and become mature (400 mg) within a month. Breeding then occurs and this is soon followed by an abrupt gonadal collapse even though day-lengths are still long. The starling has, in fact, become refractory to long days and if kept in such conditions will not show another bout of testicular growth. Normally refractoriness is broken by shortening photoperiods later in the year so that the bird can respond again the following spring. The cause of refractoriness is unknown but both prolactin and the thyroid glands are involved. Whenever refractoriness occurs in starlings in the wild, in outdoor aviaries or under experimental photoperiods, the fall in LH and FSH secretion which precedes testicular regression occurs at the same time as a rise in prolactin output. Arthur Goldsmith and Trevor Nicholls in Bristol have shown that this rise does not occur in thyroidectomized starlings and that such birds remain breeding indefinitely. However, treatment with thyroxine causes testicular collapse and, most importantly, the testes do not regrow after the treatment: the birds are truly photorefractory and must be exposed to short days before they will respond to long photoperiods.

A sidelight is that in many birds (gulls, ducks, rooks) refractoriness ends in autumn when photoperiods are quite long and, as a consequence, some reproductive activity develops. This 'autumnal sexuality' acts primarily to reinforce pair bonds, but in very favourable circumstances it can lead to nest-building and even to the hatching of young. It does not last long, however, for the days are shortening and photostimulation ceases. Not all birds show such a dramatic form of photorefractoriness and in species such as the quail, which is multi-brooded, gonadal collapse is induced primarily by the shortening day-lengths following the summer solstice. Even here though subtle effects are present which manage to terminate the breeding season prematurely. In the spring the critical day-length for quail is 12 h but, during the course of the summer, the long days slowly alter this to 15 h so that in August, when day-lengths fall below this value, the gonads regress. During the winter, short days readjust the critical day-length back to 12 h. The overall result is that quail begin to breed at the earliest possible opportunity in May but stop in late August, 2 months before day-lengths reach 12 h.

In larger mammals such as sheep, goats and deer with gestation lengths of 148–250 days, breeding takes place during the shortening days of autumn (Fig. 4.2). Like the long-day responses, this photoperiodic effect can be induced artificially, and if Suffolk ewes are exposed to LD 8:16 at the time of the summer solstice (longest day), oestrus occurs 2 months later, compared with 3.5 months when the ewes are exposed to the normal rate of daylight change outdoors (see Fig. 4.16). As long as pregnancy is prevented, ewes will continue to have oestrous cycles every 16–17 days until the end of March, but the mating season can be ended prematurely by exposing the ewes to long day-lengths. There are pronounced differences between breeds in the timing and extent of the mating season; unimproved breeds like the Soay sheep have a short season, as do some highly selected ones such as the Clun Forest, Suffolk and Welsh Mountain. Others, like the Dorset Horn and the Merino, have long breeding seasons which start early. Some temperate-zone primates are also short-day breeders, a reflection of their long gestation, so that young rhesus and Japanese macaque monkeys are invariably born in mid-summer after matings in October to December.

If the length of pregnancy approaches 1 year then spring and summer matings are once more the rule. The horse with its 11-month gestation falls into this category. Pony mares in Wisconsin show winter anoestrus and ovulate only between April and October (Fig. 4.2). Artificial selection, coupled with an excellent diet, can reduce the proportion of mares entering anoestrus in winter, but fertility in winter is still much lower than in mid-summer. By itself this would not be too important, but the racing industry has confounded the problem by arbitrarily deciding that all foals born in the Northern Hemisphere are considered to have been born on 1 January regardless of their actual birthday. Thus it is advantageous for foals to be born as early in the year as possible, and the mating season for thoroughbred mares therefore starts on 15 February. Because this is much earlier than the natural mating season, fertility is severely depressed, an expensive business when stud fees are many thousands of pounds. Exposure of mares to long photoperiods in the winter can help and some studs in France and America have adopted this treatment.

Lengthening gestation: embryonic diapause

The duration of gestation is normally fixed, but some mammals have overcome this restriction by inserting a period of embryonic diapause or delayed implantation during which the rate of blastocyst development is greatly reduced (see Book 2, Chapter 2, Second Edition). The length of this period can be either fixed (obligate delay), as in the roe deer which mates in August, implants in January and gives birth in May, or variable (facultative delay), as in the badger which mates at any time in the spring or summer, implants in December and gives birth at the end of February. Delayed implantation appears to have evolved independently in many mammals (such as kangaroos and wallabies, bears, seals, roe deer, bats, armadillos and a high proportion of mustelid carnivores), and offers advantages in various situations. In many of the seals it is necessary for mating to take place immediately after birth because the sexes live apart for the rest of the year but, since post-implantation gestation lasts for only 7–8.5 months, some form of embryonic diapause is essential to bring the total length of gestation up to 12 months. Shortening day-length appears to be the proximate cue terminating diapause in many of the seals (Fig. 4.3), but more subtle controls may also exist, as in the grey seal *Halichoerus grypus*. The mean pupping date in this species is remarkably constant within a colony, having been between 6 and 11 November for the past 26 years on the Farne Islands in the North Sea. Between colonies far more variation exists, however, and pupping dates range from mid-September (Isles of Scilly) to early March (Northern Baltic). Assuming a 'true' gestation of 249 days, the termination of diapause always occurs when natural day-lengths are increasing, but the spread between colonies seems to be too great for photoperiod to be the only proximate cue. John Coulson has proposed that sea-surface temperatures might regulate implantation, and after testing various ideas he suggests that the seals somehow add up the daily temperatures and implant when the sum reaches 800 'degree days'. All the variation between colonies can be explained if the seals begin their sums on 1 November, and he proposes that counting is activated when day-length reaches 9.5 h. The story sounds unlikely at first but temperature sums have been suggested in other contexts: all that is required is experimental proof!

Embryonic diapause is the rule amongst the macropod marsupials (and is discussed at length in Chapter 2, mainly on pages 35–44). It probably arose because the very short intra-uterine gestation is followed by an immediate post-partum oestrus. A pregnancy at this time must be held in abeyance otherwise another youngster would be produced when the first was still occupying the pouch. This potentially disadvantageous arrangement can be turned into a useful adaptation, however, in a capricious environment like the Australian desert, for it provides a reserve embryo that can be developed if the pouched youngster is lost (see Chapter 1). It is also interesting that the wallabies living in Southern Australia, like the tammar on Kangaroo Island and Bennett's wallaby in Tasmania, are photoperiodically induced seasonal breeders, in which photoperiod blocks the mother's ability to reactivate her diapausing embryo even if she loses a pouched young; by contrast, most northerly species, such as the agile wallaby, are aseasonal opportunistic breeders.

It is not immediately obvious why so many of the mustelid carnivores (skunk, badger, otter, marten, mink) show delayed implantation, but presumably it had value to their ancestors and so the phenomenon has persisted. Some mustelid species have been able to abandon it; ferrets, for example, have become typical long-day breeders with no delay and a

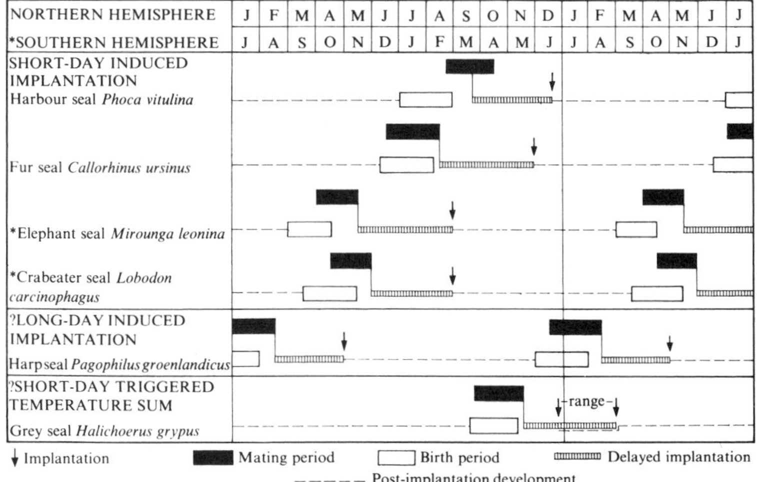

Fig. 4.3. Delayed implantation in seals. The delay often ends in the autumn or winter, perhaps by shortening days. Implantation then occurs and pups are born 7–8.5 months later. (Redrawn from D. P. Boshier. *J. Reprod. Fert.* Suppl. **29**, 143–9 (1981). Data added for the grey seal from J. C. Coulson. *J. Zool.* **194**, 563–71 (1981).)

gestation lasting for about 6 weeks. Mink usually show a brief delay of a few weeks but the time between conception and parturition is still less than 3 months. The male grows his gonads under short day-lengths and mating takes place in February/March. The lengthening days around the spring equinox are thought to break the obligate diapause by causing a resumption of progesterone secretion from the corpus luteum, thus triggering implantation of the embryo. It is still unknown how this is mediated but perhaps prolactin, increased in output as days lengthen, may act on the ovary. Many of the other mustelids have very long periods of delayed implantation, such as the 10 months of the European badger. Young are born in February/March and the females show a post-partum oestrus at which time fertilization takes place. For the next 300 days the corpus luteum is inhibited and the blastocyst grows more slowly, but in mid December implantation occurs and parturition follows 45 days later. Badgers live mostly in darkness within their setts and emerge around sunset, but by attaching radio transmitters to them Robert Canivenc in Bordeaux was able to show a close correlation between night-length and the duration of activity, suggesting that badgers monitor the seasonal shift in photoperiod and use this to time implantation. This was confirmed (Fig. 4.4) by shifting badgers to an artificial winter day-length (LD 10:14) from July to September, when all the embryos implanted within a month and

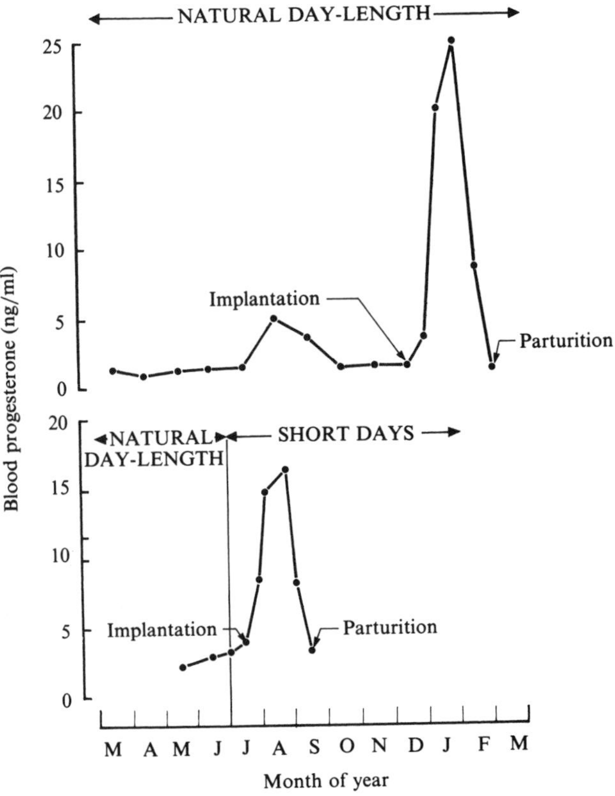

Fig. 4.4. Exposing badgers to short days in mid-summer terminates delayed implantation within 3 weeks. Progesterone rises as the corpus luteum resumes functioning and young are born 2 months later. The natural cycle is shown for comparison. (Redrawn from R. Canivenc and M. Bonnin. *J. Reprod. Fert.* Suppl. **29**, 25–33 (1981).)

were born in September. These examples represent just two of the variants. Many martens mate in July/August and carry an unimplanted embryo until the spring, when long days then activate the corpus luteum and birth takes place 30–50 days later. In these animals short days can extend delayed implantation, while long days terminate it prematurely.

Photoneuroendocrine machinery

Viewed simply, photoperiodism requires three components. First, there must be a photoreceptor to detect light, and a 'clock' to enable it to distinguish 'long' from 'short' day-lengths. Secondly, there must be neural pathways linking the clock to the neuroendocrine apparatus. Probably these are the components that can be modified or overridden by other environmental factors. Finally, there is the endocrine system itself, involving pituitary gonadotrophins, gonadal development and gonadal steroid feedback. In practice it has been possible to investigate these components only in the relatively few species that can be kept successfully in the laboratory: Japanese quail, white-crowned and house sparrows, mallards, starlings, hamsters, voles, ferrets, mink, wallabies, sheep, deer and, most recently, the horse.

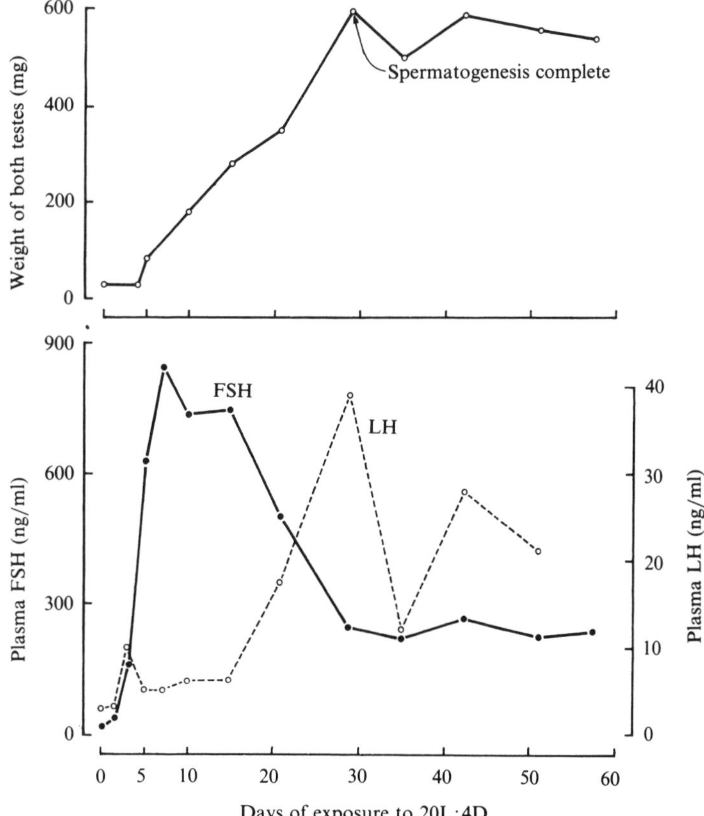

Fig. 4.5. Photostimulation of Djungarian hamsters with long days triggers gonadotrophin secretion. In turn, this causes spermatogenesis and sexual maturity within 4–6 weeks. (Redrawn from S. M. Simpson, B. K. Follett and D. H. Ellis. *J. Reprod. Fert.* **66**, 243–50 (1982).)

Gonadotrophin secretion. The primary response to photostimulation is increased pituitary secretion of LH and FSH. This is particularly obvious in male animals, the testes of which grow enormously over a few weeks; Fig. 4.5 shows data from the Djungarian hamster. The output of FSH increases 40-fold within 10 days of a change in photoperiod and this stimulates spermatogenesis. Gross changes in circulating LH are less obvious with infrequent blood sampling, but androgens rise rapidly, causing the FSH levels to decline somewhat. Spermatozoa appear after a month and the hamster is fully fertile within 6 weeks. A similar pattern of hormone secretion is also seen in birds on long days and in rams on short days. The sequence of changes is similar to that seen at puberty in rats and men. Although the time-scales may be different between species, the similarities suggest that the endocrinology of gonadal development is probably common to all: what differs is the neural mechanism used to activate the hypothalamo-pituitary complex.

Minute-by-minute analyses of LH and FSH produce a slightly different picture, with LH in particular oscillating with remarkable regularity (Fig. 4.6). From this has developed the idea that the hypothalamus contains a 'pulse generator' which fires regularly to cause a co-ordinated discharge of GnRH from the median eminence (see Book 3, Chapter 1, Second

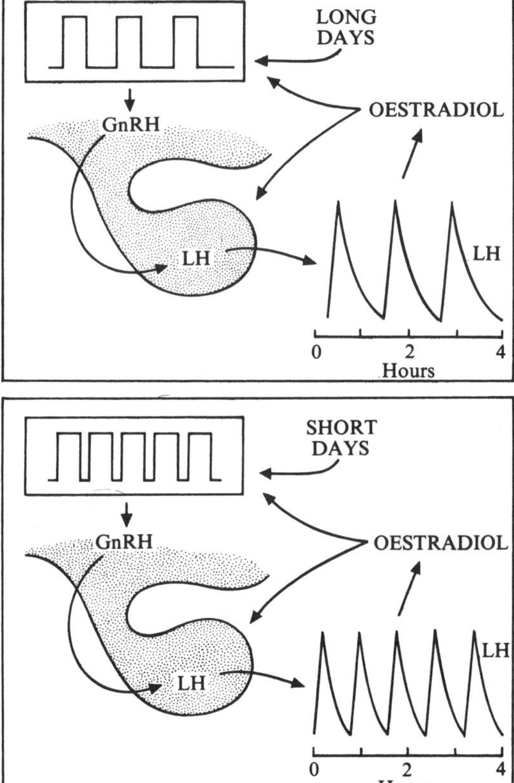

Fig. 4.6. A model to show how photoperiod may increase gonadotrophin secretion by altering the activity of the hypothalamic pulse generator and hence regulating GnRH secretion. (Adapted from R. L. Goodman and F. J. Karsch. In *Biological Clocks in Seasonal Reproductive Cycles*, pp. 223–36. Ed. B. K. Follett and D. E. Follett. Colston Papers No. 32. Wright Scientechnica; Bristol (1981).)

Edition). Photoperiod may act, therefore, by altering the frequency of the pulse generator, the best evidence for this having been found in sheep in which the daily number of LH secretory episodes rises two- to three-fold between summer and winter.

Photoreception and the clock. Mammals use their eyes as the photoreceptors for measurement of day-length, but the pathways are unusual. Information flows along the optic nerves in neurones separate from the visual system (the retinohypothalamic tract) and terminate in the suprachiasmatic nuclei, two discrete regions located in the hypothalamus just above the optic chiasma. It is notable that birds do not rely solely on their eyes to detect photoperiodic light, since they also have extra-retinal photoreceptors which detect light diffusing through the skull. The exact location of these photoreceptors in the hypothalamus is unknown, but an action spectrum shows them to be rhodopsin based. People often express surprise when they hear that birds possess such non-visual receptors, but we should remember that both the eyes and the pineal organ (a third photoreceptor in sub-mammalian vertebrates) derive embryologically from the third ventricle, as does the hypothalamus.

The timing device seems to be based upon a circadian (approximately 24-h) 'clock', rather than upon some kind of interval timer, and is located in the suprachiasmatic nuclei. These small areas, each containing a few thousand neurones, regulate all the body's circadian rhythms; destroying them upsets rhythms of sleep–wakefulness, adrenal function and temperature change. The photoperiodic clock also ceases to function after such lesions, so that a hamster, for example, is left in a permanent long-day situation, unable to regress its gonads under short days. Much careful thought about how circadian rhythms could be used as a photoperiodic clock has produced many ideas but perhaps the simplest is to imagine that, as day-lengths alter, so the output from the central clock changes its characteristics and this conveys the message about day-length. For instance, under short days the signals from the photoperiodic clock may not be inductive, but on exposure to long days the phase and/or duration of the signal may alter to become inductive. This idea is not without foundation, of course, and is consistent with much that is known about the ways other circadian rhythms respond to light–dark cycles.

A prediction that follows from the concept of a circadian-based clock is that induction depends upon the timing of photic events and not on their duration. It should occur if the photoperiod is simulated by two short light pulses, one acting as 'dawn', the other as 'dusk', and if the gap between the pulses is varied it should be possible to determine how far apart they must be before the animal reads the period as a 'long day'. This experiment has been successful in a number of species. For example, if the second light pulse falls 11–15 h after the first 'dawn' pulse, quail (Fig. 4.7) react as if the day is long and grow their gonads accordingly. Any other combination is read as a short day. The duration of the pulses can be very brief indeed, and some experiments in Bristol by David Ellis have shown that two pulses of only 1 s in duration, set 14 h apart, cause gonadal growth in golden hamsters. Even shorter pulses (250 ms) will maintain mature hamsters in breeding condition, a clear demonstration that it is not the duration of light each day but only in when it occurs that is important.

Neural/endocrine links. The pathways in birds seem to be relatively straightforward in the sense that both the photoreceptor and the clock probably lie within the hypothalamus. Information then flows to the GnRH cells bodies in the anterior hypothalamus, thereby regulating the anterior pituitary gland. A very different route exists in mammals since they use the pineal as the principal neuroendocrine transducer. This has been discussed by Gerald Lincoln in Book 3, Chapter 3, Second Edition, but a brief resumé is in order (Fig. 4.8).

The first point to make is that most of our information has emanated

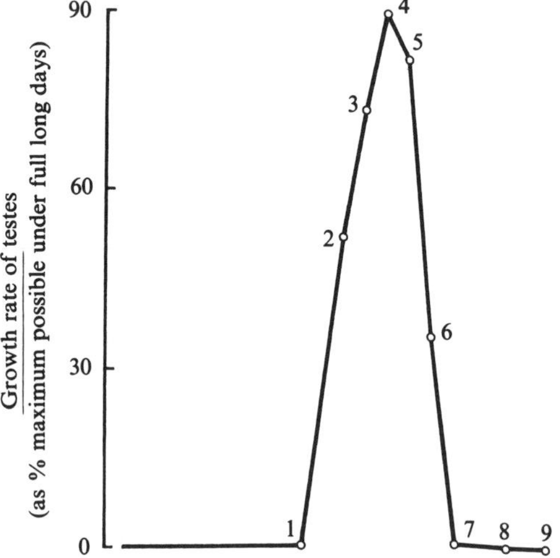

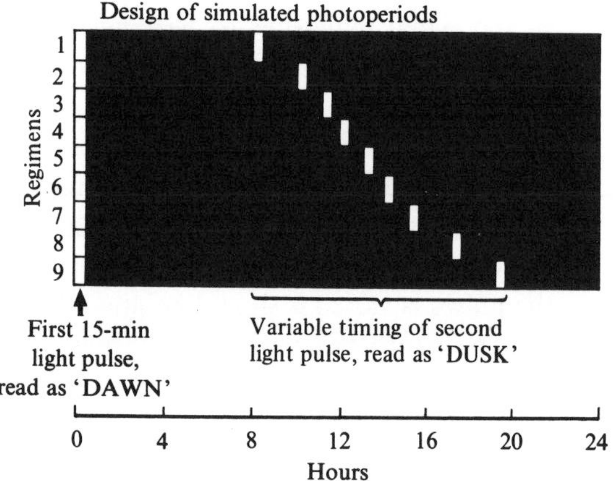

Fig. 4.7. The photoperiodic clock. Because day-length is measured by a circadian clock, light need not be present all the time for the bird or mammal to think the day is 'long', and this experiment shows how two 15-min light pulses per day can trigger testicular growth in quail kept in darkness. Not all possible arrangements, of course, are interpreted as long days and the curve traces out those times in the 24 h when the bird is sensitive to light: the circadian rhythm of photoinducibility. The maximum testicular response is seen under lighting regimen No. 4, when the two pulses are 12 h apart. (B. K. Follett and P. J. Sharp, unpublished data.)

from a small number of laboratories in America and Germany where hamsters have been studied. Only recently has the work been extended to sheep and to other species. After the photic information has been processed in the suprachiasmatic nuclei, nervous impulses pass through the brain stem and down the spinal cord. Preganglionic sympathetic nerves leave the thoracic part of the cord and ascend the neck to synapse in the superior cervical ganglia. From here postganglionic sympathetic nerves travel to the head to innervate many structures, including the pineal gland. It is only when the pineal is stimulated by the sympathetic nervous system that melatonin is released into the circulation. Normally this occurs only at night since pineal activity is regulated by a circadian clock lying in the suprachiasmatic nuclei. That this complex pathway is crucial for seasonality came as a surprise, but the effect in a golden hamster of removing the pineal, or of severing its sympathetic innervation, is to render the animal no longer photoperiodic and it remains reproductively active regardless of day-length. Somehow, therefore, the pineal gland can turn off the reproductive system of hamsters under short days. Since melatonin injections in the evening but not in the morning cause testicular regression, it is likely that the daily pattern of melatonin secretion carries the photoperiodic message. In effect, the evening injections extend the period of natural melatonin secretion to simulate that occurring in short-day conditions. This hypothesis is supported by the pattern of pineal melatonin secretion in Djungarian hamsters; it differs greatly between short and long day-lengths.

Bruce Goldman and his colleagues have carried the matter further by infusing melatonin into pinealectomized hamsters for 4–6 h each day, so mimicking a long-day pattern of secretion, or for 10–12 h as occurs in short days. Testicular regression takes place only in the latter case. The animal,

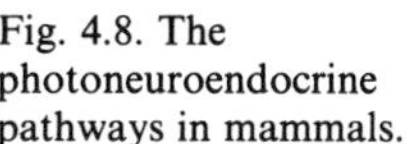

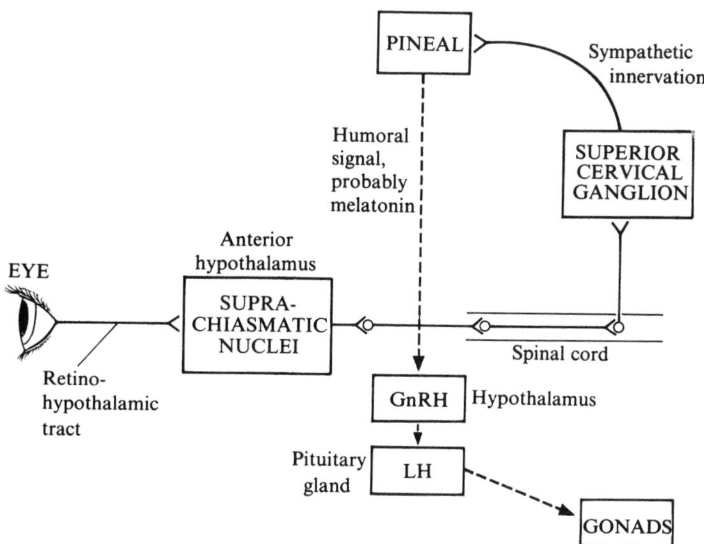

Fig. 4.8. The photoneuroendocrine pathways in mammals.

therefore, possesses a calendar in the form of the profile of melatonin secretion, and this is somehow interpreted by the hypothalamus in an appropriate fashion. The word 'appropriate' is used because the pineal is not associated with seasonal reproduction alone but also with a range of other photoperiodic changes. These include the increased secretion of prolactin and increases in food intake in the long days of summer and the changes in hair growth and colour in hamsters, mink and sheep as a result of short days.

To see melatonin as simply an 'antigonadotrophic' hormone because of its inhibitory effects in hamsters is too narrow a concept. How a species uses the melatonin information depends upon its particular requirements, and this point becomes clear if we consider the sheep, a species breeding in short days. Pinealectomy abolishes the animal's ability to respond to photoperiod, and Eric Bittman has been able to re-establish seasonal oestrous cycles in such animals by infusing melatonin for 16 h per day, whereas infusions for only 8 h switch off LH secretion and produce anoestrus. Pineal denervation also abolishes seasonal prolactin secretin in sheep, deer and goats, abolishes seasonal embryonic diapause in the tammar wallaby, and upsets breeding in horses.

Clearly major unresolved questions remain. Is the photoperiodic message enshrined only in the duration of the nightly release of melatonin or does it also involve the amplitude of the rhythm and its phase position relative to dawn? Where is melatonin acting within the hypothalamus and how does the brain unravel the message and respond? In a sheep, for example, long-day information triggers prolactin secretion, but short-day information increases gonadotrophin output. Also, why do birds not use the pineal gland (or melatonin) to regulate seasonal breeding? The gland exists and conveys photoperiodic information just as in mammals, but pinealectomy has little or no effect on any aspect of the avian photoperiodic response.

Circannual reproductive cycles

By now the reader must feel that in the absence of the appropriate photoperiodic stimulus, reproduction becomes aseasonal. In many cases this is true, for example white-crowned sparrows held on short day-lengths never grow their gonads, whereas Japanese quail on long days breed continuously. But some animals can continue to have annual reproductive cycles even when kept in constant photoperiodic conditions. Such endogenous annual rhythms were first shown 25 years ago in hibernating golden-mantled ground squirrels *Spermophilus alteralis*. Under constant light or constant darkness or uniform day-lengths (LD 12:12), the squirrels' annual 5 months of winter hibernation did not disappear, but persisted over 4 successive years, although the periodicity of the rhythm (i.e. the time from the onset of one hibernation to the next) differed significantly from 365 days. In a total of 61 measurements, the average was 341 days (range 229–445 days), so that by the fourth year many squirrels were entering hibernation 3–4 months earlier than the controls under natural illumination. Such results are in accord with the concept of an endogenous clock in the squirrel which shows its intrinsic periodicity under conditions of constant photoperiod. In normal life, this clock is entrained to exactly 365 days by some environmental cue, most probably the photoperiod. By analogy with circadian rhythms, these longer-term oscillators were called circannual clocks. Subsequently they were found in other species, notably in garden warblers and blackcaps which winter in equatorial Africa but breed in central Europe. When kept in constant photoperiod, moulting and migratory restlessness recurred every 10 months, and individuals showed nine complete cycles over 8 years. In each cycle there were, of course, two attempted bouts of migration, one to and one from the wintering grounds, and even this was reflected in the circannual clock. When tested in an orientation cage, during the 'autumnal' period the warblers tended to head south, but during the 'spring' migratory period they headed north-west. The adaptive significance of evolving such long-term clocks is reasonably obvious in both these cases: squirrels hibernating in their burrows lack the opportunity to use a direct environmental cue to tell them when to emerge, and warblers living on the equator are in much the same situation.

Migration is certainly associated with reproduction in birds but do circannual clocks regulate gonadal function? There are numerous accounts of annual breeding cycles in sheep persisting under constant photoperiods (even in pinealectomized sheep!), but these cycles may well be driven by pheromonal, nutritional or temperature signals; it is very difficult to design

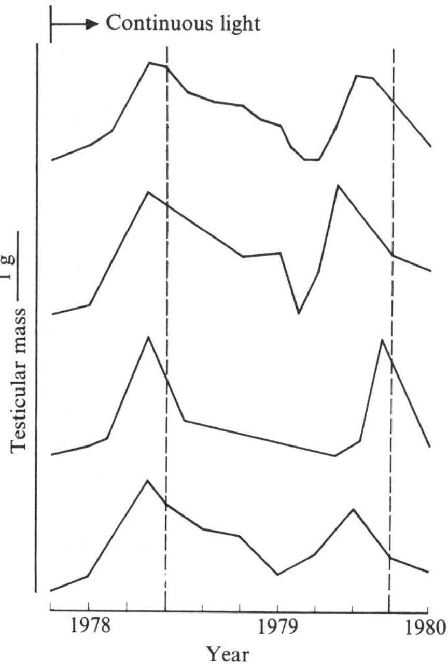

Fig. 4.9. Circannual reproductive clocks. Antelope ground squirrels continue to show rhythmic patterns of testicular growth even in constant environmental conditions, but the period length is not 365 days, indicating the endogenous nature of the clock. Bar on the ordinate represents the scale for 1 g in weight. (Redrawn from G. J. Kenagy. *J. Comp. Physiol.* **142**, 251–8 (1981).)

a crucial test of endogenous (circannual) rhythms in mammals. Richard Goss subjected sika deer to a variety of constant photoperiods and, though numbers were small, cycles of antler growth and shedding, both of which are testosterone-dependent, persisted with a 10-month periodicity for up to 3 years. More convincingly, Jim Kenagy found that testicular size in golden-mantled and antelope ground squirrels kept in constant conditions changed as much as in the wild and persisted with a periodicity of about 300 days (Fig. 4.9). This would be essential for the golden-mantled squirrel which lives in a mountain environment where the breeding season is severely constrained by snow-cover, so that there is only just enough time to squeeze in gestation, lactation and the post-weaning growth periods before the snows return. Fertilization must take place within 2 weeks of breaking hibernation and this means that testicular development must be initiated during hibernation. Analogous problems exist for the desert-living antelope ground squirrel which does not hibernate but still grows its testes during the winter because the productivity of the desert prior to the summer drought demands breeding in the very early spring.

Nutrition, rainfall and temperature

Despite nutrition being central to reproduction, its importance is invariably underplayed. Food availability around the time of birth not only determines when young are produced, but also influences the number of young surviving to puberty. Great tits are so dependent upon caterpillars that the number of eggs laid is regulated by caterpillar availability just before egg-laying, and since caterpillar emergence is itself controlled by the warmth of the spring, temperature underlies the nutritional effect. Mammals show analogous responses. Nutrition in the later stages of pregnancy is important for fetal growth, and lambs born to ewes fed on a low plane of nutrition weigh 40 per cent less than normal. Similarly, poor nutrition affects lactation and hence the development of the young, especially if they are born in a very immature condition. The average mouse litter, for example, is 45 per cent of the mother's weight at birth, but 2 weeks later it is 150 per cent, and without copious food the mother cannot sustain lactation. In grey seals, the young are suckled intensively for 2 weeks while the females starve and lose 40 kg in weight. Half of this goes to the young who more than double their birth weight, the rest representing energy used by the mother. In such conditions survival to weaning is closely correlated with the mother's nutritional state and hence her milk supply. Nutrition, therefore, determines reproductive success, but can it act as the proximate cue for the onset of reproductive activity? Probably food availability, or the rainfall that determines it, does regulate the timing of reproduction in many tropical species but, apart from strong correlations, there is no direct evidence. Conception in giraffes is closely linked to when the acacia are at their best, and rainfall–nutrition correlations have also been suggested for the timing of conception in elephant and springbok, and the red kangaroos living in central Australia.

Clearer cases exist in rodents since they are particularly susceptible to nutrition, being rapidly disadvantaged by food shortages but also able to take advantage of a temporary superabundance. This is seen most dramatically in deserts. When Janice Beatley catalogued fluctuations in rainfall and plant populations in the Mojave desert in relation to rodent numbers (primarily kangaroo rats, *Dipodomys* spp.) she concluded that reproductive timing was strongly influenced by the animals' water balance, which in turn depended upon the succulent green food produced after the winter rains. This effect has also been shown experimentally in pocket mice *Perognathus formosus*. Animals held on short days and fed lettuce daily had a mean testicular mass of 56 mg, but on long days and with the same quantity of lettuce the testes increased to 140 mg; controls without green food hardly grew their testes at all (10 mg). There is obviously a subtle dependency of small iteroparous rodents on a range of environmental variables. One mechanism may normally be dominant but may be overridden; thus even in short days and low temperatures natural populations of house mice fed extra wheatgrass will start winter breeding (Fig. 4.10). Improved nutrition can also accelerate puberty (see below). Presumably the variables interact with one another; cold temperatures accelerate gonadal regression in short days, whereas warmth and food can maintain the gonads despite the days being short. None of this surprises ecologists but it can be confusing to physiologists as they try to untangle the routes through which these variables act.

Temperature was once thought to be responsible for seasonal breeding but, in fact, it is difficult to find instances where it plays a major role. One of the few is provided by the heterothermic bats which vary their body temperature. Paul Racey in Aberdeen was able to prolong gestation in

Fig. 4.10. In feral house mice food availability determines seasonal breeding. In this experiment, carried out near San Francisco, the percentage of lactating females dropped virtually to zero between December and April but feeding grain in December to one group overcame the seasonal decline and the mice regained reproductive activity. (Redrawn from K. T. Delong. *Ecology*, **48**, 611–34 (1967).)

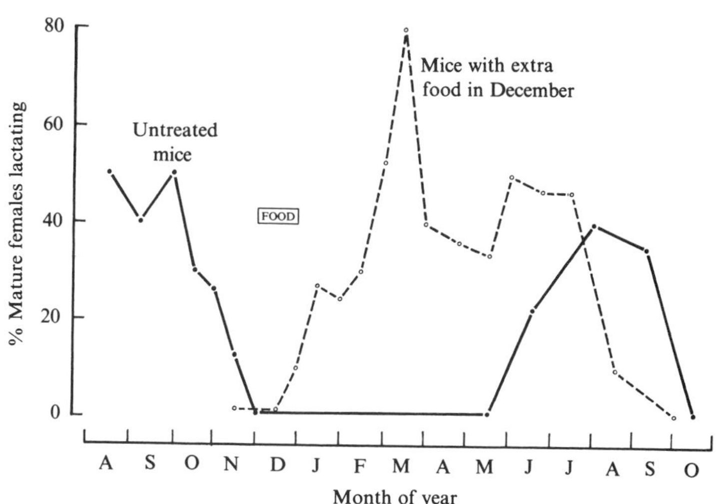

pipistrelle bats by a combination of low temperature and food deprivation, a response also occurring in the wild under harsh conditions. Near Aberdeen a colony gave birth 10 days later in 1979 than in 1978, most of this difference being explicable by a period of cold, wet weather in mid-pregnancy (May–June) during which the bats went into torpor. Pipistrelles also exhibit the other adaptation of delayed fertilization. The males grow their testes in June to July and inseminate the females in the autumn. The spermatozoa remain in crypts within the uterus for 6 months until ovulation occurs in mid-May when fertilization takes place, and the young are born 40–50 days later.

A specific effect of heat occurs in male mammals because high temperatures impair spermatogenesis. This has been known for a long time, because cryptorchids, animals whose testes do not descend into the scrotum, are invariably infertile. Heat loss from the scrotum normally keeps the temperature 2–4 °C cooler than in the body cavity but in extreme environmental conditions this difference can be obliterated. A well-attested case is the summer sterility affecting rams in Central Australia, but it also occurs in natural populations of wild animals that have evolved to live in such high temperatures. Alan Newsome found a correlation between the percentage of male red kangaroos *Megaleia rufa* with spermatogenetic defects and the number of exceedingly hot (> 40 °C) days. Up to 80 per cent were badly affected and this seemed to account for the zero pregnancy rates during hot droughts.

Finally, hot humid weather depresses coital rates in man and is the cause of seasonal variation in conceptions in many parts of Asia (as for example in Hong Kong, see Fig. 6.20 in Book 3, Second Edition). The effect, however, is relatively minor, amounting to a depression of about 20 per cent compared with the mean birth rate.

Social environment

Once the discussion turns from the coarse adjustment mechanisms such as photoperiod or nutrition to the fine tuning mechanisms that are essential for mating and conception, then the diversity of social signals used, and the complexity of their interactions, make it difficult to obtain a clear overview. Social signals enable one individual to tell another of the same species about its presence and its inclinations, but they may also tell the recipient something about the sender's dominance and the quality of his territory. All this can be encapsulated in the vocalizations that males make to attract females and deter other males; examples abound, from the roaring of a red deer stag or a howler monkey, through the posturing and trumpeting of a male elephant seal, to the degree of song complexity in a bird. The types of signal used encompass all the sensory systems – visual, auditory, olfactory and tactile – and the combination used reflects the sensory make-up of the particular species. Olfaction and ultrasound are important in rodents, whereas visual stimuli predominate in fish, birds and terrestrial primates. If a bizarre sensory system has evolved, it too may be adopted, and nothing illustrates this better than the electrical communication systems developed by various mormyrid and gymnotoid fish living in the murky waters of west African rivers. Signals are based upon the patterns of discharges emitted from electric organs, usually in the tail, and are analysed by the recipient using skin electroreceptors derived from the lateral line system. Carl Hopkins has shown that these signals can be used for sex recognition. Male fish produce a triphasic discharge longer than that of the female and, in addition, the discharges are produced in patterns. One pattern is a 'rasp' with 20 to 30 discharges delivered quickly over 50 ms and followed by silence for several hundred milliseconds. Rasps are heard only during the breeding season from males in the presence of females, and are not produced by isolated males or juvenile fish. Fig. 4.11 summarizes an experiment in which the discharges from a captive fish were played through the water to a male living in the river. Those from a female trigger a response greater than those from another male or from a female of a closely related sub-species. Species- and sex-recognition by some kind of signal is widespread in every group of animals.

Few studies have attempted to evaluate the significance of social cues in the breeding biology of the species in its natural habitat; the most compete data come from rodents, mainly the house mouse *Mus musculus*. As one of the great opportunists it has become commensal with man and inhabits dwellings and food stores from the tropics to as far north as the Lofoten Islands at 69 °N. Populations can be found living at −10 °C in cold stores, or in farm buildings where the mean temperatures are always above 28 °C. Feral mice range across 116° of latitude and inhabit many different climates, although their origin on the Steppes of Russia makes them especially abundant in more arid regions and much less so in

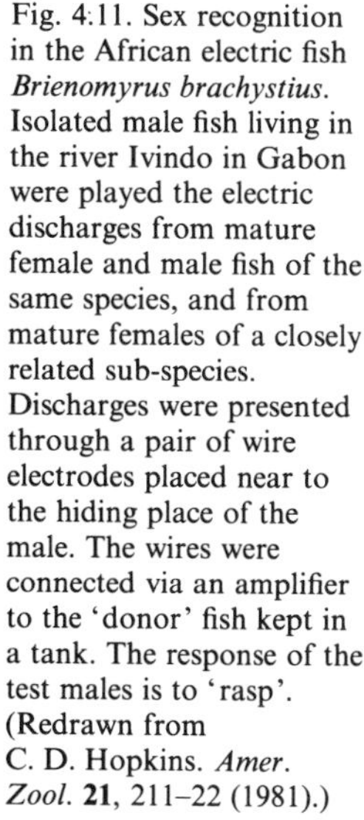

Fig. 4:11. Sex recognition in the African electric fish *Brienomyrus brachystius*. Isolated male fish living in the river Ivindo in Gabon were played the electric discharges from mature female and male fish of the same species, and from mature females of a closely related sub-species. Discharges were presented through a pair of wire electrodes placed near to the hiding place of the male. The wires were connected via an amplifier to the 'donor' fish kept in a tank. The response of the test males is to 'rasp'. (Redrawn from C. D. Hopkins. *Amer. Zool.* **21**, 211–22 (1981).)

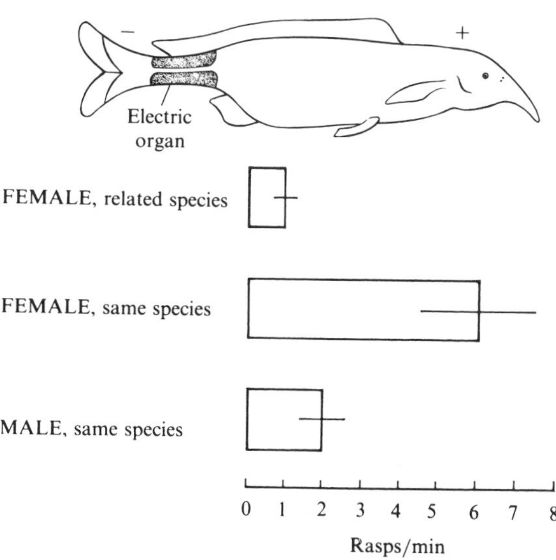

woodlands and tropical jungle. (The distinction between feral and commensal populations is often difficult to draw, since in harsh conditions feral animals exist only during the warmer parts of the year, being replenished annually from the overwintering commensals.) In stable environments with a plentiful food supply population densities of commensal mice are high, ranging up to 10 mice per square metre. The territories are typically less than 10 m² and are dominated by a single male who lives with several breeding females, some of their offspring and a few subordinate males. Little gene flow takes place between territories, and population growth is by dispersal. In contrast, densities of feral mice are much lower, with possibly one mouse per square metre, and the home ranges are large (1000 m²). Territorial behaviour may exist, but the sheer size of the territories and the instability of the populations (mortality may be 30 per cent per month and overwintering losses 90 per cent) suggest that it is less rigidly organized than in high-density commensal situations. Seasonal breeding is a characteristic of many feral populations, with food not photoperiod being the primary cue, but commensals often breed throughout the year.

Once the population is established, social cues of an olfactory nature (pheromones) take over (see also Book 3, Chapter 6, Second Edition). Mice produce a range of pheromones which are excreted in the urine. Their chemistry is still not known but they are certainly capable of acting downwind, and must be volatile or form an aerosol. Males are the primary sources of pheromones and they routinely mark the ground with urine. By the simple expedient of putting filter paper on the floor of the cage and subsequently viewing this in ultraviolet light, Frank Bronson and his colleagues showed that males mark as often as 500 times per hour, far more frequently than females. This marking tells other males of the territorial boundaries but it also alters the female's reproductive state directly.

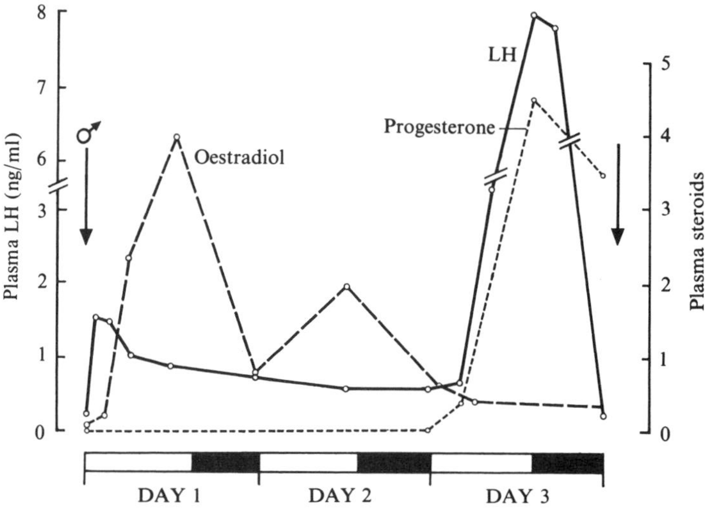

Fig. 4.12. Changes in hormone secretion in pre-pubertal female mice exposed to an adult male (♂). The immediate response is rapid and the females go through puberty and ovulate within 72 h. (Units for oestradiol are pg/ml, and for progesterone ng/ml.) (Redrawn from F. H. Bronson. *Q. Rev. Biol.* **54**, 265–99 (1979).)

The onset of puberty is accelerated in females (Whitten effect), and when no adult male is present puberty is postponed and oestrous cycles, when they do occur, are prolonged and fitful. As little as 10 μl of male urine per day, delivered as an aerosol, can trigger puberty by stimulating pituitary gonadotrophin and gonadal steroid secretion (Fig. 4.12). Within 1–3 h of exposing young females to an adult male, levels of LH rise and this increases plasma oestradiol 15-fold. Subsequently, a classic pre-ovulatory sequence develops with oestradiol triggering an LH surge and ovulation. Perhaps a separate pheromone acts on mature females for if they are kept isolated from a male their oestrous cycles become disorganized. Male urine can synchronize them, however (Lee–Boot effect), again by increasing LH secretion and suppressing elevated prolactin levels. The production of these male priming pheromones is totally dependent on androgens and thus control is exerted primarily, if not solely, by the dominant male in the territory. The females are not entirely passive though and produce a urinary pheromone that triggers LH release in males. Thus the female can stimulate the male to stimulate the female!

Finally, there are pheromones that suppress reproduction. Females can counteract the acceleratory effects of the male in inducing puberty and can also slow down the onset of puberty in other females by 3 weeks or more. The particular chemosignal responsible is found in the urine of both single-caged and grouped females, but in contrast with most male pheromones gonadal steroids are not essential for its production, although it disappears after adrenalectomy. Another signal from males blocks implantation in pregnant females and causes them to return to oestrus (Bruce effect). It is not difficult to imagine advantages for both the inhibitory and acceleratory actions should individuals of either sex find themselves in a new territory and, while the primary function of the mouse's pheromonal cueing system is to produce mating, it can also accelerate population growth at low densities or hold it in check when numbers are high (Fig. 4.13). Some fascinating field evidence for this has been found in isolated populations of wild house mice by Adrianne Massey and John Vandenbergh who showed that urine collected from the females living in high density

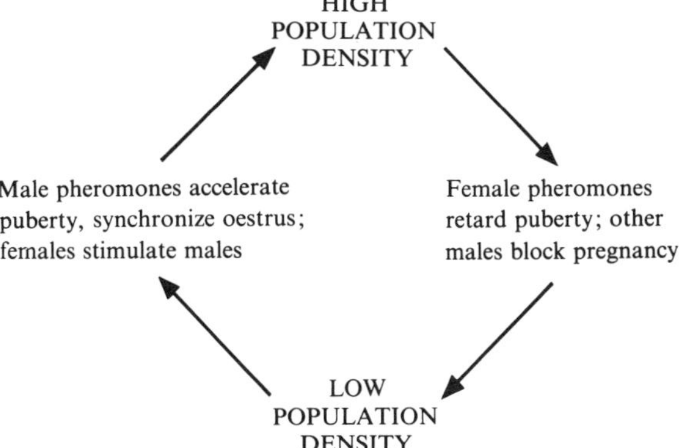

Fig. 4.13. A model of how populations of rodents might be regulated by pheromonal cues.

was able to delay puberty in laboratory females but that urine from low-density populations did not affect puberty.

Rodents may use ultrasound (>20 kHz) as the synchronizing agent for the act of mating itself. Gonadal hormones regulate the level of ultrasound production and its wave form, and in addition there is a strong inter-sexual effect which is triggered by pheromones. Male hamsters, for example, increase their rates of calling many-fold after coming into contact with a female and detecting her vaginal pheromone; this effect is even greater if the female is in oestrus. The net result is that ultrasound controls the social behaviour preceding mating. A female hamster enters a male's territory because of her proceptive behaviour when in oestrus, at which point the male picks up her vaginal odour and is stimulated to produce ultrasound. This increases the female's own emission of ultrasound and the animals approach one another. Fertilization is facilitated by the male who calls at a high rate throughout copulation, thereby prolonging lordosis in the female and allowing him an opportunity for multiple ejaculations. When it is remembered that hamsters in the wild are solitary animals living almost permanently underground, the significance of these ultrasonic signals can be fully appreciated.

Stress at the height of the breeding season must be considerable. The classical response to stress is to increase adrenocortical activity, and in woodchucks (*Marmot monax*) the weights of the adrenal glands do increase during reproduction. An even more dramatic instance is in various dasyurid marsupial shrews (*Antechinus* spp.) where all the males die after mating! This occurs because a stress-induced increase in circulating corticosteroids coincides with high androgens, which depress the hepatic production of corticosteroid-binding proteins. The effect is to elevate free corticosteroids to pathological levels and to cause death by lowering disease resistance and triggering inflammatory responses and haemorrhage in the upper digestive tract.

Man as an environmental factor

Man influences reproduction in other mammals by domestication, by exploiting wild populations and so altering their population density, by changing their habitat, and by pollution. The general effects are well known, but only rarely is there persuasive evidence of a direct effect on reproduction (as distinct from an indirect one through mortality) and it is a bitter irony that the best case comes from man's desire to catalogue in detail the characteristics of every whale he has slaughtered. The annual catches of the four main baleen whales in Antarctica since 1910 are shown in Fig. 4.14. Initially, blue and humpbacked whales were taken until their low numbers made hunting unrewarding, and so attention turned to the fins, then to the sei whale and now to the minke whale. Estimates of the stocks prior to exploitation show the blue whale to have decreased from 180000 to 8000, fin whales from 400000 to 85000 and sei whales from

150000 to 50000. Such reductions have had a substantial impact on the amount of krill (*Euphasia* spp.) available and Richard Laws estimates its consumption to have fallen from 270 to 130 million tonnes per year. The 'excess' of 140 million tonnes per year has in turn benefited the remaining whales and altered their reproduction in various ways, one of the most impressive being a reduction in the age of puberty (Fig. 4.15). Fin whales in the thirties reached puberty at 9 or 10 years of age, but within a decade

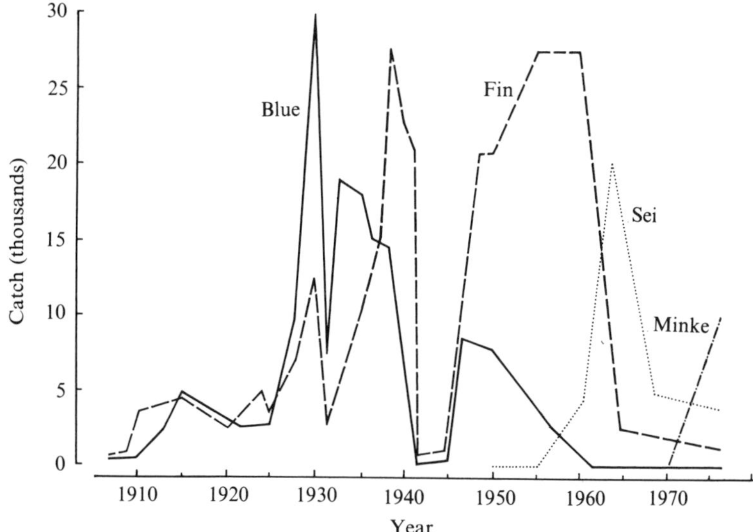

Fig. 4.14. Annual catches of baleen whales in the Antarctic. (Redrawn from R. Gambell. *J. Reprod. Fert.* Suppl. **19**, 533–53 (1973).)

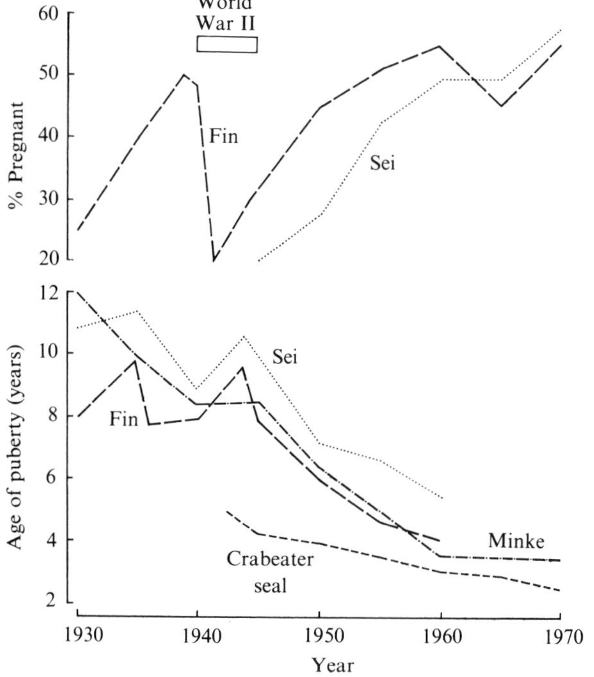

Fig. 4.15. The evidence for changes in pregnancy rates of adult females, and age at puberty, in whales and crabeater seals over the past 50 years. (Redrawn in the main from R. M. Laws. *Phil. Trans. Roy. Soc. B*, **279**, 81–96 (1977).)

this had fallen to 5 or 6 years. This is a remarkable change, even more impressive than the 5-year decline in the age of menarche in girls over the past century, and it means that fin whales are now growing at an average rate of 4 m per year. This has not been the only change, however, for another response has been an increase in pregnancy rates. Until 1928 no lactating fin whale that was also pregnant had been caught, but soon after that up to 20 per cent of lactating females were carrying a fetus. These changes have occurred very rapidly. Pregnancy rates declined as fishing ceased in World War II and krill once more became limited, but rates had increased again by 1950. The main cause of the change is that the whales have shortened their normal 2-year breeding cycle (1 year gestation, 6 months lactation, 6 months rest) to 1 year, the females becoming pregnant again immediately after giving birth. Initially, these effects on puberty and ovulation were thought to be a direct response to the falling population density but since fertility also increased in the (then) non-exploited minke and sei whales, as well as in crabeater seals, the reason would seem to be the increased availability of krill.

Agriculture, another of man's major activities, has also altered reproduction in animals. Much of this has been due to selection for fertility, and the improved nutrition of his farm animals, but he has also made some aseasonal by optimizing photoperiod and temperature. Egg production in poultry is a classic example: the winter period of ovarian regression can be avoided completely by keeping the days long. The modern egg-laying strains of chickens are not especially photoperiodic, but the margin between profit and loss is so small that even a minor effect becomes important. Thus by using a photoperiodic treatment regimen which involves 4 months of short days (LD 6:18) from the day of hatching, and then a weekly increase of 20 min to LD 17:7, egg production begins a few days earlier than under less advantageous schedules. Turkeys pose a greater problem because such a high proportion develop broodiness and become photorefractory. This problem could perhaps be solved by using marginally stimulatory day-lengths long enough to grow the gonads but too short to cause refractoriness, or by tampering with thyroid function (see p. 107).

Of the larger farm animals only sheep, goats and horses show strong reproductive seasonality. The involvement of the pineal in regulating this (see pp. 114–16), and the discovery by Larry Tamarkin that properly timed injections of melatonin to hamsters alter the photoperiodic response, suggested a method of altering the breeding season of the sheep. By feeding melatonin-soaked pellets in the late afternoon to ewes in mid-summer, the breeding season can be advanced by many weeks (Fig. 4.16). The effect occurs because the profile of melatonin in the bloodstream is altered to that seen in short days.

Even pigs and cows, however, do show minor seasonal trends in reproductive efficiency that could be worth removing for commercial reasons. In one long-term study on a highly efficient pig farm, small increases in the time of return to oestrus after farrowing, in the abortion rate and in the incidence of infertility were found during the later summer months. Overall, the fall in production arising from this was only 3–5 per cent but nation-wide it amounts to a loss of £10–20 million per year. High temperatures, acting as a stressor, and possible photoperiod (the wild pigs in the Chizé forest in France first start coming into oestrus in December and are all in anoestrus from May until September) are both involved.

Even greater commercial rewards could come from controlling seasonal events other than those strictly linked to reproduction. For example, growth rates in many ruminants (sheep, red deer, reindeer, cattle) are higher in summer than in winter, a fact that at present farmers have to accept. However, it now appears that photoperiod may lie behind this change for it also regulates appetite: the voluntary intake of food is greater in long days. There are hints that the pineal may be the transducer in this process also and, if this is so, methods might be found to alter the secretion of melatonin during the winter months, thus leading the animal's appetite centre to imagine that it is still summer! The alternative, of course, is to expose animals to long day-lengths, but this is feasible only where cattle or sheep are brought into central wintering quarters. This often happens in Michigan, and Allen Tucker has increased heifer growth rates by 10 per cent with extra illumination at night. Another important seasonal change is in lactation. The interrelationships between milk production, temperature, quality of pasture and prolactin secretion in dairy cows are complex, but the low levels of prolactin in winter certainly do not help milk yields. Since plasma prolactin rises during long days, Allen Tucker exposed two matched groups of Holstein cows, which calved in mid-September, to

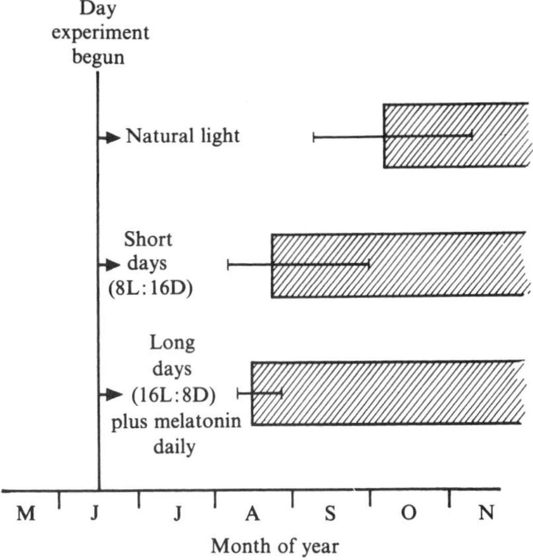

Fig. 4.16. Beginning of oestrous cycles in three groups of five Suffolk ewes. In mid-June one group was moved to short day-lengths (8L:16D), another left in natural daylight and the third given long days (16L:8D) together with a pellet soaked with 3 mg melatonin daily at 16.30 h (8 h after dawn). This extra melatonin, acting in concert with the normal nocturnal secretion from the sheep's own pineal gland, was just as effective as a short day, and advanced oestrus by 50 days. (Taken from J. Arendt, A. M. Symons, C. A. Laud and S. J. Pryde. *J. Endocr.* **97**, 395–400 (1983).)

either natural (decreasing) day-lengths or to LD 16:8, giving all the cows the same amount of food. Over the next 100 days (Fig. 4.17) the photostimulated group consistently produced 10–15 per cent more milk. The economics are clear and, with modern management procedures, perhaps floodlighting could be used to improve production, especially since the light need be given only for 2 h in the middle of the night to stimulate prolactin secretion (cf. Fig. 4.7).

The environment and man's reproduction

Despite a good many speculations to the contrary, man is not grossly affected by factors such as day-length, but observations by Ehrenkranz (1983) show that Eskimo women do have a peak of conception in June, to give a peak of births the following March. The seasonal variation which exists in births in the Anglo-Saxon world seems to owe much more to the incidence of holiday periods than to anything else. In many warmer parts of the world the seasonal changes are greater, and as mentioned previously, in the tropics conceptions tend to be inversely correlated with temperature (see Fig. 6.20 in Book 3, Second Edition).

There is a long history relating nutrition to human fertility. Famine is invariably associated with temporary falls in the birth rate, a fact known since medieval times although the best documented case was the 50 per cent fall in conceptions during the Dutch famine in the winter of 1944–45; a similar sequence of events was seen 30 years later during the rice famine in Bangladesh. Self-inflicted starvation can also stop menstruation, as is seen in women suffering from anorexia nervosa. A less serious situation develops in women who, as athletes or ballet dancers, involve themselves in severe exercise (Fig. 4.18). Running serves as a good example, and with

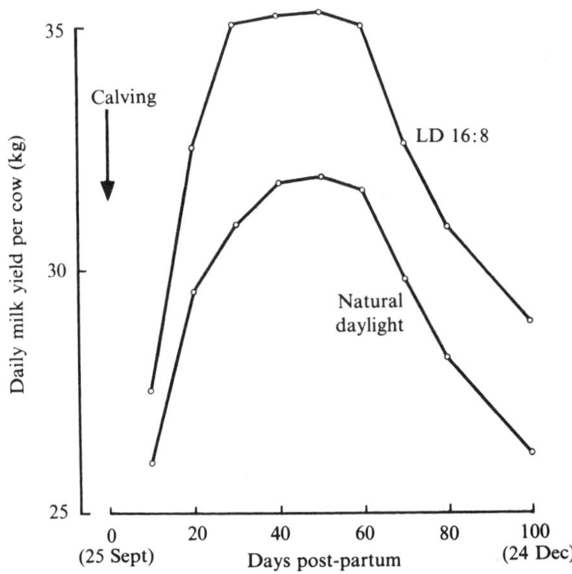

Fig. 4.17. Milk yields during the autumn and winter in post-partum Holstein cows given the same amount of food and kept on natural day-lengths (Michigan) or on LD 16:8. (Redrawn from H. A. Tucker and W. D. Oxender. *Prog. Reprod. Biol.* **5**, 155–80 (1980).)

a weekly mileage of 50 miles or more over a third of women become amenorrhoeic. The same problems arise with ballet dancers, swimmers, and even female recruits to West Point Military Academy. There is much argument as to whether weight loss is the primary cause of this amenorrhoea. Body fat is certainly lower in amenorrhoeic runners (17 per cent) than in those training for more than 30 miles per week but still ovulating regularly (21 per cent), and this difference also extends to daily protein and to weight loss after the onset of running, with the amenorrhoeic women losing twice as much weight. This argues in favour of Rose Frisch's general hypothesis of a minimum weight being necessary for continuing menstrual cycles, but other data are not consistent with this concept. Michelle Warren, for instance, found in her study of young ballet dancers that injuries preventing exercise were followed by a resumption of menses without a change in body

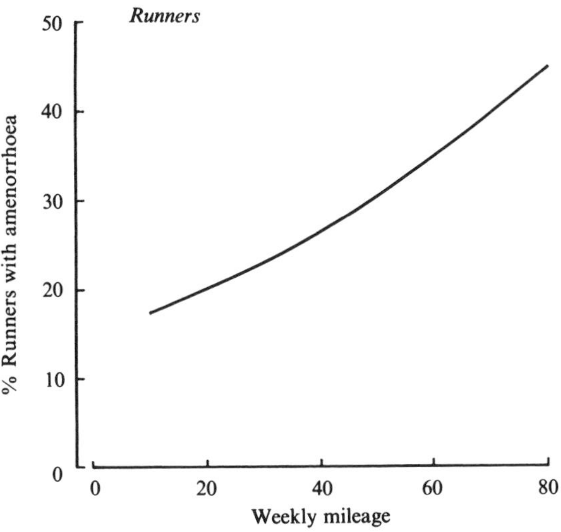

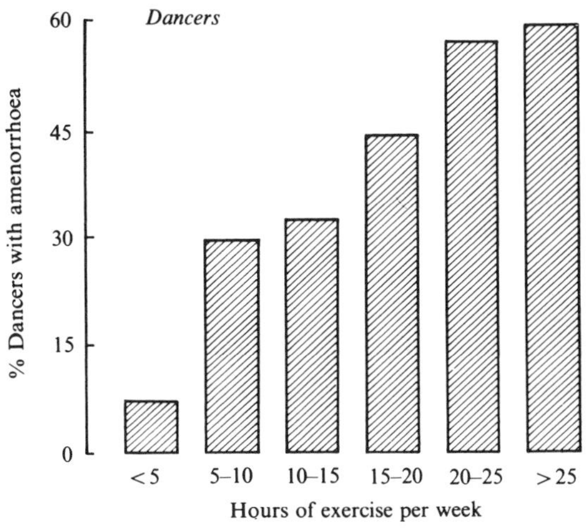

Fig. 4.18. The amount of running in athletes and of exercise in young ballet dancers as related to the incidence of amenorrhoea. (Redrawn from C. B. Feicht, T. S. Johnson, B. J. Martin, B. E. Sparkes and W. W. Wagner. *Lancet*, **2**, 1145–6 (1978). Additional data from M. P. Warren. *J. Clin. Endocr. Metab.* **51**, 1150–7 (1980).)

weight. Stress may also be a major factor and when this is combined with weight restriction amenorrhoea is precipitated.

A more pernicious influence on human reproduction is exerted by long-term socio-economic circumstances. Rose Frisch has argued that family size in the lower social classes of mid-Victorian Britain was constrained by poor nutrition, hard physical work and appalling living conditions; together these produced a short and less efficient reproductive span. The effects are actually worse, however, because the health and physique of the mother, often a direct reflection of her own nutrition during infancy and childhood, is a determinant of her child's health. Sir Dugald Baird has provided the evidence for this association; in Aberdeen between 1948 and 1952 the perinatal mortality rate in children born of women of poor physique was 62.8 per 1000, compared with 26.9 per 1000 for those of good physique. A significant difference also occurred in the percentage of babies weighing less than 2.5 kg at birth (a critical weight) and was still present in their ultimate heights when full-grown: 13 per cent of the children from mothers of poor physique were taller than 162 cm compared with 42 per cent from very healthy mothers.

Nutrition is also involved in the timing of puberty (see also Book 3, Chapter 6, Second Edition). For the past century the age of menarche has steadily diminished in most western societies and now occurs at about 13 years. This is 3 years earlier than in Bangladesh and nearly 6 years earlier than in the Bundi tribe living in the New Guinea highlands. These differences are generally attributed to nutrition but exactly how this influences reproductive physiology is much argued.

The last 20 years have amply demonstrated just how all-pervasive the environment is in regulating reproduction. We shall doubtless continue to catalogue more effects and will emerge with many fascinating examples, the significance of which may be baffling. Is it true, for instance, that the stress suffered by many German women in war-time pregnancies led to alterations in the degree of fetal androgenization and so to the greater-than-average percentage of homosexual males claimed for this cohort? Can it really be true that Californian quail regulate their breeding by sex-steroid feedback, using not their own steroids but phytoestrogens they derive from eating the desert plants? Certainly during very arid seasons the phyto-oestrogen content rises, and in those years the quail show irregular nesting. Fascinating as such ideas are they must not dominate our thinking, for there are some central physiological and ecological questions that must be answered.

The first of these relates to the neural routes whereby each environmental factor regulates gonadal function. It is attractive to imagine that a final common pathway exists, perhaps involving the hypothalamic pulse generator and the GnRH neurosecretory neurones, and that this generator receives a host of positive and negative inputs from higher neural circuits. Long day-lengths, for example, are stimulatory to a hamster but are counterbalanced by inhibitory influences ranging from sex steroids to nutritional deprivation; the final effect is the sum of these various influences. We need not only to know the precise neural pathways involved in transmitting information about photoperiod and/or steroid feedback, but also a means of quantifying the 'drive' from each component. If the latter could be measured then answers to questions such as whether seasonality is really regulated by hypothalamic changes in sex-steroid feedback sensitivity might be forthcoming. Some of us view this as a prime effect of photostimulation whilst others see it only as a secondary and inevitable result of increasing direct photoperiodic drive. Understanding photoperiodic time measurement is a clock problem that is of central importance to our understanding of seasonal breeding. We should not lose sight of the fact that there may be a constellation of physiological events controlled by photoperiod, such as food appetence, metabolic activity, ability to respond to hormonal stimulation, each of which has an indirect effect on reproduction.

A second set of questions relates to the relative importance of the environmental variables in a given species. The very layout of this chapter tends to over-emphasize the importance of particular factors, such as photoperiod or nutrition or pheromones, and could lead to the impression that species respond to a limited number of environmental inputs. In reality, of course, they respond to all inputs, although the combination will be different depending upon the reproductive strategy.

Perhaps one of the greater challenges that lies ahead for the environmental physiologist is to determine precisely how environmental factors regulate the natural rates of population growth of species in the wild, since this is the key to their survival.

Suggested further reading

Symposium on 'Social signals – comparative and endocrine aspects'. *American Zoologist*, **21**, 111–316 (1981).

Environment and reproduction. D. Baird. *British Journal of Obstetrics and Gynaecology*, **87**, 1057–67 (1980).

Pineal melatonin secretion drives the reproductive response to daylength in the ewe. E. L. Bittman, R. J. Dempsey and F. J. Karsch. *Endocrinology*, **113**, 2276–83 (1983).

Seasonal breeding in humans: birth records of the Labrador eskimo. J. R. L. Ehrenkranz. *Fertility and Sterility* **40**, 485–9 (1983).

The reproductive ecology of the house mouse. F. H. Bronson. *Quarterly Review of Biology*, **54**, 265–99 (1979).

Photoperiodism and seasonal breeding in birds and mammals. B. K. Follett. In *Control of Ovulation*. Ed. D. B. Crighton, N. B. Haynes, G. R. Foxcroft and G. E. Lamming. Butterworths; London (1978).

Latitude of origin influences photoperiodic control of reproduction of deer mice (*Peromyscus maniculatus*). J. Dark, P. G. Johnson, M. Healy and I. Zucker. *Biology of Reproduction*, **28**, 213–20 (1983).

Biological Clocks in Seasonal Reproductive Cycles. Ed. B. K. Follett and

D. E. Follett. Colston Papers No. 32. Wright Scientechnica; Bristol (1981).

Exercise and reproductive function in women. D. C. Cumming and R. W. Lebar. *American Journal of Industrial Medicine*, **4**, 113–25 (1983).

Population, food intake and fertility. R. E. Frisch. *Science*, **199**, 22–30 (1978).

Environmental Factors in Mammalian Reproduction. Ed. D. Gilmore and B. Cook. Macmillan; London (1981).

Controlled Breeding in Farm Animals. I. Gordon. Pergamon Press; Oxford (1983).

Seasonal reproduction: a saga of reversible fertility. F. J. Karsch. *The Physiologist*, **23**, 29–38 (1980).

Effect of melatonin feeding on serum prolactin and gonadotrophin levels and the onset of seasonal estrous cyclicity in sheep. D. J. Kennaway, T. A. Gilmore and R. F. Seamark. *Endocrinology*, **110**, 1766–72 (1982).

Seals and whales of the southern ocean. R. M. Laws. *Philosophical Transactions of the Royal Society, London, B*, **279**, 81–96 (1977).

Seasonal breeding, Nature's contraceptive. G. A. Lincoln and R. V. Short. *Recent Progress in Hormone Research*, **36**, 1–52 (1980).

Photoperiodism and Reproduction in Vertebrates. Ed. R. Ortavant, J. Pelletier and J.-P. Ravault. INRA Colloquium 6. INRA; Paris (1981).

The pineal and its hormones in the control of reproduction in mammals. R. J. Reiter. *Endocrine Reviews*, **1**, 109–31 (1980).

Seasonal reproduction in higher vertebrates. Ed. R. J. Reiter and B. K. Follett. *Progress in Reproductive Biology*, **5**, 1–221. Karger; Basle (1980).

Pheromonal regulation of puberty. J. G. Vandenbergh. In *Pheromones and Reproduction in Mammals*. Ed. J. G. Vandenbergh. Academic Press; New York (1983).

5

Reproductive behaviour

E. B. KEVERNE

In recent years our understanding of reproductive behaviour has expanded to include not only the underlying physiological and endocrine mechanisms, but also the mating strategies that animals adopt in their natural environment. A survey of animal mating patterns reveals a remarkable variety in the frequency and form of mate selection. While some animals take but a single mate in their lifetime, others may acquire many, either successively or simultaneously. Whatever strategy is adopted, it is assumed that this is a balance between the costs and benefits, the benefits being the increased contribution to the gene pool, and the costs being a failure to thrive or survive.

The mating system characteristic of a species is a fundamental aspect of its social organization, and has developed in response to selection in the animal's ecological niche. While the ecology may account for the kind of differences we observe between the species, the different strategies adopted between males and females within a species are probably related to the amount of investment they make in their offspring. Darwin was aware of the competition between males for access to breeding females and described how sexual selection resulted in masculine traits that enhanced an individual's ability to attract females (inter-sexual selection, sometimes called 'epigamic') or defeat competitors (intra-sexual selection). Several studies have shown that the reproductive success of male mammals depends on intra-sexual selection and is closely related to their fighting ability, as determined by their strength, or body size, or the development of their weapons, such as horns or antlers. In contrast, females depend on inter-sexual selection, and exercise more care in choosing a mate and in raising their offspring.

Parental investment

In 1972 Robert Trivers formulated a general theory of parental investment, which he defined as any investment in offspring that increased the chances of the offspring's survival, at the cost of the parent's ability to invest in other offspring. In the case of mammals it is invariably the female that invests more heavily in each offspring, since the energy costs of gestation and lactation are immeasurably greater than those of spermatogenesis and copulation.

In theory, the reproductive success of a monogamous pair of animals will be the same for each of them. Consequently, the parent making the greater energy investment per offspring will be the limiting resource. The male could increase his reproductive success by opting for polygyny; this would also increase the degree of sexual selection, resulting in the development of secondary sexual characteristics that arise as a result of intra-sexual selection pressures.

The Trivers model can also be applied to parental investment through time; at any moment there will be a temptation for the partner with less accumulated parental investment to desert the other. This is particularly true for the male immediately following insemination, when the female's investment is on the increase. If the female is deserted, she may still be able to maintain her pregnancy but fail to produce viable offspring because the task is overwhelming. In this case the deserting male pays the price of an ultimate loss in his own genetic fitness. If such a loss was not compensated for by increased success in future matings with other partners, then the selective pressures would be against desertion, thereby favouring monogamy.

Mating strategies in the wild

Monogamy

In a monogamous system each male mates with only one female for life and vice versa (Fig. 5.1). Although common in birds, this type of mating

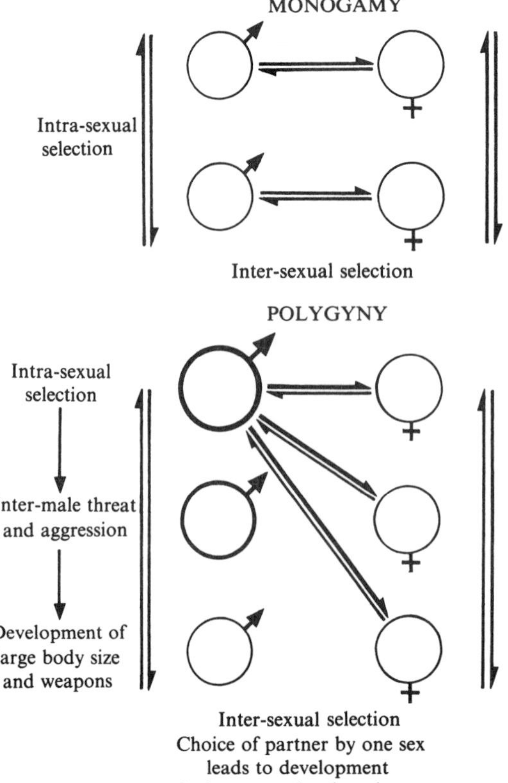

Fig. 5.1. The inter- and intra-sexual relations in monogamous and polygynous systems. (From R. V. Short. *Advances in the Study of Behavior*, **9**, 131 (1979).)

strategy is rare in mammals. It will evolve only when the need for co-operation in rearing offspring outweighs the advantage to either partner of seeking extra mates. Such conditions are most likely to arise in environments that are poor in valuable resources. Nevertheless, monogamy is among the more highly evolved forms of social organization since it often involves a considerable degree of tolerance towards the partner over a long period of time and often outside the context of mating.

In a monogamous system, because the male is investing so much in the care of the offspring, it is genetically advantageous to him to make sure that he has exclusive access to the female's unfertilized eggs, and is not 'cuckolded'. This he does by becoming extremely territorial and possessive of his female so as to prevent adultery. Since the intra-sexual selection pressures are no greater between males than they are between females, there is an absence of marked sexual dimorphism. The formation of the pair bond (courtship) is also a long and protracted affair; since both sexes are investing all their eggs in one basket, so to speak, then they must make certain it's a worthwhile basket! Although courtship may be long, monogamy is associated with infrequent sexual behaviour (except in man) and low rates of reproduction. In wild canids and beavers, sexual behaviour is restricted to a single short period each year (monoestrous). The females must conceive at that time or wait a whole year before they can become reproductively active again. Sexual behaviour is therefore relatively intense over several days and may include a genital lock which 'ties' the pair together for up to 30 min (Fig. 5.2). The female acouchi (a South American rodent) comes into oestrus approximately every 40 days (polyoestrous), but with a gestation period of 3.5 months and lactational anoestrus of 7 weeks, mating may actually occur only twice a year. Among the monogamous primates, sexual activity is rarely seen in the siamang and other gibbons, while long-established breeding pairs of tamarins and marmosets engage in sexual activity infrequently even in the laboratory. It therefore seems unlikely that sexual behaviour is a major determinant of pair-bond maintenance in monogamous species, except in human society.

Fig. 5.2. Dogs *in copula*, in the distinctive tail-to-tail position.

Monogamy is also characterized by a relatively long period of maturation for the young, which are usually with their parents throughout this time. In marmosets, tamarins, gibbons, jackals, dwarf mongooses and beavers, the young may often aid in the care of the next litter and thus create a family unit. Within such family units the young may be inhibited from reproducing. This is seen particularly in wolf packs which are composed of related individuals, but where only the dominant parents breed. Related wolves in the pack which do not breed but may help 'altruistically' in the care of siblings are said to perpetuate their own genes by a process of kin selection.

Two forms of monogamy have been described. Facultative monogamy may result when a species exists at very low densities, with males and females being so spaced that only a single member of the opposite sex is available for mating. Obligate monogamy appears to occur in a species when a solitary female is unable to rear her litter successfully without help from conspecifics, but the carrying capacity of the habitat is insufficient to allow more than one female to breed simultaneously within the same home range. In both forms the adult male aids in rearing the offspring, especially in those monogamous species that produce large litters (e.g. canids) where the male may participate in feeding both the mother and her offspring.

Polygamy

In polygamous species an individual has more than one mate, either successively or simultaneously (Fig. 5.1). The most common form is polygyny, in which one male mates with a number of females. In polyandry, one female mates with two or more males, while in promiscuous species there are no pair bonds and both males and females mate with more than one member of the opposite sex.

Polyandry has not been observed among mammals, except in certain human societies. This is understandable in view of the excessive investment made by the mother during her pregnancy and lactation. However, in a small percentage of avian species, where the male has assumed the full burden of incubating the eggs and rearing the brood, the female has been emancipated from maternal commitments and has the potential for increasing her biological fitness by production of multiple clutches. In the few known cases of avian polyandry there is intense female–female competition for males, and females are the larger, more aggressive sex. This suggests that the differences in secondary sexual characteristics observed between the sexes are not necessarily inherent aspects of masculinity, but may be viewed as a consequence of sex differences in reproductive strategies. Even in mammals, there are numerous exceptions where females are larger than males, in bats, whales, hyenas and hamsters, for example.

Polygyny occurs when the environmental or behavioural conditions bring about the clumping of females, and males have the potential to monopolize them. This is biologically adaptive in large social groups where males can mate with several females without any dire consequences for their sperm reserves. Polygynous societies are characterized by distinct sexual dimorphism brought about by intra-sexual selection pressures, and epigamic displays brought about by inter-sexual selection pressures. Males tend to be larger than females and more aggressive, a feature aided by the development of weaponry (antlers, horns, large canine teeth) which have the triple role of competing with males, displaying to females and protecting females and infants as a group from predators. One consequence of male competitiveness may be the formation of dominance hierarchies, although these may also be present among females. Females of polygynous species tend to develop epigamic displays to attract males (sexual swellings and sexual colouration together with postural displays).

The strategy adopted by males to monopolize females in polygynous species may be expressed in a number of different forms. Males may compete among themselves for access to resources that are essential to females. Habitats that have an uneven distribution of food supply result in a mosaic of male territories of different qualities. For example, female hummingbirds incubate and rear their young alone, but in order to do so require an abundant and reliable source of nectar. Males allow females nesting in their territory to feed therein, but aggressively exclude all other hummingbirds.

Males may acquire a 'harem', a form of polygyny in which males aggressively herd females and exclude other males from the area. Such behaviour is seen in many ungulate species, where the females and young aggregate into small groups. During times of impending sexual receptivity these groups may be herded by the males who will attempt to defend a moving territory around as large a group of females as they can muster.

A third type of polygyny is one in which males defend neither females nor a territory rich in some food resource, but rather determine among themselves who is the boss and how they rank in relation to other group members. This form of male dominance polygyny will depend upon the degree of synchronization of sexual activity among females of the population. When females are highly synchronized and converge upon males within a short period of time, a highly promiscuous 'explosive' breeding situation will occur. If the females of a population are relatively asynchronous in their periods of sexual receptivity, the males remain active for the duration of the breeding season, and this results in the establishment of a stable dominance hierarchy, or lek.

Sexual behaviour

No study of reproduction can be considered complete without its taking account of sexual behaviour. This not only involves coitus, when the male gametes are transferred to the female, but also the bringing together in courtship of males and females at a time appropriate to optimize fertile matings. Such synchronization of sexual behaviour with other components of reproduction is determined by the blood level of gonadal hormones. Just as the genital tract is a target organ for the gonadal hormones, so too is the central nervous system, and the behaviour of the animal serves as indicator of the response of the brain to humoral stimulation. (This topic is also discussed in Book 2, Chapter 3, second edition.)

Sexual behaviour appears only periodically in the life of most female animals and is triggered when the plasma concentration of gonadal hormones has risen to a critical level. The signal for this upsurge in hormonal secretion is frequently environmental – light, nutrition, odours, tactile stimulation, social stimulation, temperature – and the effects of environmental changes are transmitted via the central nervous system. Thus, while the CNS is under the influence of gonadal hormones for the expression of behaviour, the release of gonadal hormones is itself regulated via the CNS. It is tempting to promote the brain or the gonad to the rank of conductor in this sexual orchestra. It can be argued that the CNS has both the first and final say in the expression of sexual behaviour, but in most species it can exercise this dominance only if the endocrine environment is appropriate. Even in higher primates and man, where sex is predominantly a cerebral affair, the brain is still not totally emancipated from the influence of the gonadal hormones.

Gonadal hormones and sexual behaviour

In considering the action of gonadal hormones on sexual behaviour it is helpful to separate the peripheral effects, that is those acting on target tissue other than the brain, from those acting centrally on the brain. Of course this separation does occur naturally when the testes or ovaries secrete their hormones into the blood, but in order to evaluate the relative contribution of peripheral and central effects of hormones on behaviour, such distinctions are essential. It is important to consider not only where hormones are active, but when in life they have their action. In the adult they usually have either a facilitative or an inhibitory action on behaviour, but perinatally they have an important inductive action on differentiation of target tissue.

Early production of gonadal hormones and their role in sexual differentiation

We have recognized for several decades that the secretions of the male gonad exert a somatic effect on the rudimentary genital tract in the fetus, but it was only during the decade 1960–70 that a proliferation of experiments demonstrated, in a number of species, a critical time in the perinatal period when gonadal hormones also exert a considerable influence on the sexually undifferentiated brain. The stimulus for these studies arose from experiments by William Young in the United States in which it was revealed that testosterone given to pregnant guinea-pigs produced female offspring that, on reaching sexual maturity, responded to exogenous androgen by showing mounting behaviour similar to that normal in males.

Contemporary experiments by Geoffrey Harris in the United Kingdom revealed similar effects in rats, and showed that their critical period for sexual differentiation of the brain was the fourth day after birth. A single injection of testosterone to female pups at this time is sufficient to depress their feminine sexual behaviour later, but enhance male patterns of behaviour if these rats are ovariectomized and given testosterone when adult. Conversely, castration of male pups during this critical period enhances the display of feminine patterns of copulatory behaviour in adults in response to exogenous ovarian hormones, but also permits male patterns of sexual behaviour if exogenous testosterone is administered instead. Experiments such as these led to the conclusion that late in embryonic life, or even soon after birth, the central nervous system is sexually undifferentiated and the future patterns of sexual behaviour are organized by the secretions of the immature testes. In addition to the effect on rodents (rat, mouse, hamster, guinea-pig), exposure of females to perinatal androgen has now been shown to enhance the masculine response potential of the dog, ferret, sheep, the marmoset monkey and the rhesus monkey.

Somewhat paradoxically, we now believe that the testicular androgens are not themselves responsible for masculinization of the brain, but require first to be converted by aromatization to oestrogen. Female fetuses may be protected from the masculinization effects of their own and maternal oestrogens by a high rate of oestrogen metabolism to less active components, and subsequent conjugation, as well as by the protective powers of an oestrogen-binding protein (α-fetoprotein) found in CSF. This binding protein has a high affinity for oestrogen but little for testosterone. It is reasonable to suppose that the amount of freely available oestrogen reaching the relevant brain sites of females is very low, whereas the unbound testosterone in males is readily available for aromatization to oestrogen in the appropriate neurones, thus ensuring masculinization.

Experiments such as these provide a 'biological' basis for explaining sexually dimorphic behaviours, and it is therefore tempting to relate such findings to our own species. Cross-species extrapolation to man is, however, fraught with dangers, but these become obvious only when experiments with closer related primate species are made together with an examination of the clinical literature. Thus, in the rhesus monkey, administration of testosterone to pregnant females results in female offspring that have masculinized external genitalia, and since they still possess ovaries, they are classed as pseudohermaphrodites. This fetal masculinization does indeed produce subsequent sex differences in behaviour that are more similar to those shown by males than females. Moreover, these behavioural differences appear in the first 3 years of life, long after the testosterone has disappeared, and yet long before sexual maturity. They are expressed as 'rough and tumble play' and 'immature (or 'foot-clasp') mounting' (Fig. 5.3). However, when these androgenized female monkeys reach maturity some show male mounting behaviour when paired with females, and female patterns of sexual behaviour when paired with males. Only a small percentage of these monkeys ever copulate, but this is probably more a reflection of their limited social experiences during infancy. However, they do start normal ovulatory menstrual cycles after puberty, regardless of the extent of external masculinization. Early social experiences themselves produce major effects on the development of sexual behaviour in primates, ranging from complete sexual inadequacy as a result of early separation from the mother and subsequent social deprivation, to differences in 'rough and tumble' play behaviour and mounting, when raised with the same- or different-sexed peers. The fact that raising male and female monkeys separately enhances early mounting in females and suppresses this behaviour in males is itself an indication of some of the difficulties encountered in interpreting the effects of hormones on the ontogeny of sexuality in primates.

Evidence for sexual differentiation of the brain in humans comes from

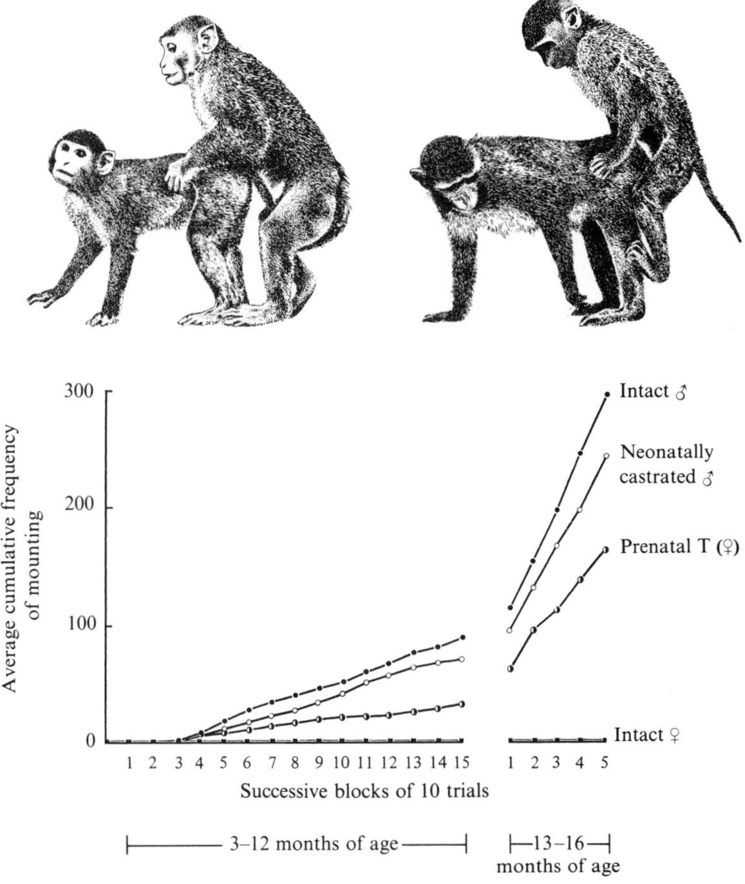

Fig. 5.3. Foot-clasp mounting by an androgenized female rhesus monkey (right), contrasted with that by a normal female (left), who stands on the floor for the purpose. The graph illustrates that the development of this immature form of mounting is much higher in females given prenatal testosterone (T) than in non-treated (intact) females, but is lower than in normal males or in neonatally castrated males. (From R. W. Goy. In *Recent Advances in Primatology*, pp. 449–62. Academic Press; New York (1978).)

clinical studies, especially on the adrenogenital syndrome where exposure of the fetus to androgen occurs as a result of a pathological overactivity of the adrenal glands. Another group of androgenized human subjects has appeared as a result of the injudicious use of synthetic progestagens given to mothers to prevent recurrent abortions. It transpired that these synthetic progestagens have androgenic properties which masculinize female offspring, and such therapy was thereafter abandoned. Both these groups produce female babies that appear with masculinized external genitalia, and until recently were raised as boys (Fig. 5.4). Since the 1950s, the appropriate cortisol therapy for adrenogenital syndrome has prevented postnatal over-production of androgens, and the early recognition of the syndrome has not resulted in such mistakes in gender assignment; clinically, this group most closely resembles the experimental group of monkeys that were androgenized *in utero*. Behaviourally also they are very similar; as children they show more energy expenditure in play ('tomboys') but when they become adults there is little evidence to suggest that their sexual behaviour is any different from control female subjects matched for age and social class. Their psychosexual orientation is that of a female; they prefer men, get married and have children. Clearly then, early exposure to high androgen levels *in utero* does not have irreversible consequences on the brain sex of these subjects.

The same is not true for girls with untreated adrenogenital syndrome; they are reared as boys and continue to be exposed to very high levels of testosterone into puberty. These girls have a male appearance, sexually prefer girls, and usually wish to abandon their genetic sex and be formally re-assigned as males, even though they have no testes and a comparatively small 'penis' (enlarged clitoris). Sexual re-assignment involves ovariectomy and testosterone replacement therapy which must then continue throughout life. Somewhere between these groups (those with the syndrome treated or untreated) is a condition that was first reported for an isolated

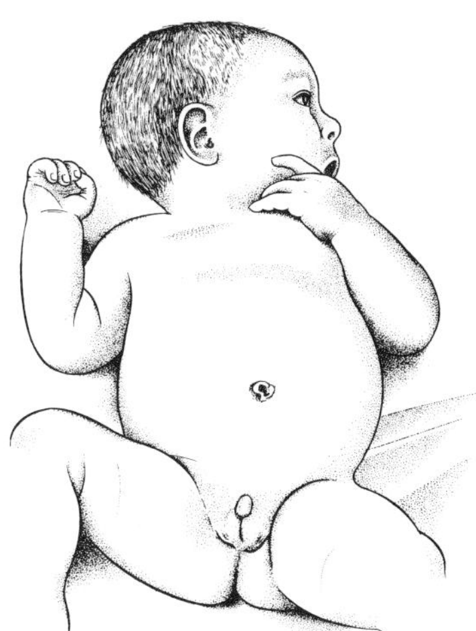

Fig. 5.4. Typical appearance of the external genitalia of a baby girl born with hyperactive adrenals (adrenogenital syndrome), or whose mother received progestagen treatment during pregnancy. (From J. L. Hamson and J. G. Hamson. In *Sex and Internal Secretions*, pp. 1401–32. Ed. W. C. Young. Baillière, Tindall & Cox; London (1961).)

population of families in the Dominican Republic. A deficiency in a recessive gene carried by the females results in a deficiency of the enzyme 5α-reductase, which produces a condition in males whereby insufficient amounts of dihydrotestosterone are metabolized from testosterone. As a consequence the external genitalia are not fully masculinized at birth, and boys (*guevodoces*) with this inherited condition are reared as girls. At puberty, the increased production of testosterone is sufficient to overcome the enzyme deficiency and masculinization proceeds, together with a re-assignment of psychosexual identity to that of a male. At face value, this finding would suggest that early exposure of the brain to testosterone appears to overcome the sex of rearing. Such a view is, however, not without its critics since the gender identity of these boys is not entirely unambiguous. Their collective name '*guevodoces*' (literally translated as 'eggs at 12') implies a special place for these children in the eyes of the community, and they themselves are aware of their differences by the age of 6–7.

The clinical equivalent of early castration in males can be found in testicular feminization which occurs in genetic males with physiologically normal levels of circulatory androgens but with target tissues that are insensitive to these androgens (Fig. 5.5). The subjects have genitalia that are unambiguously female at birth and they are subsequently reared as girls. At puberty, masculinization fails to occur, and these genetic males develop breasts and have all the appearance of normal women except that they have scant pubic hair. Psychosexually, their orientation is directed towards men, they often marry, and although infertile show normal maternal behaviour towards adopted children.

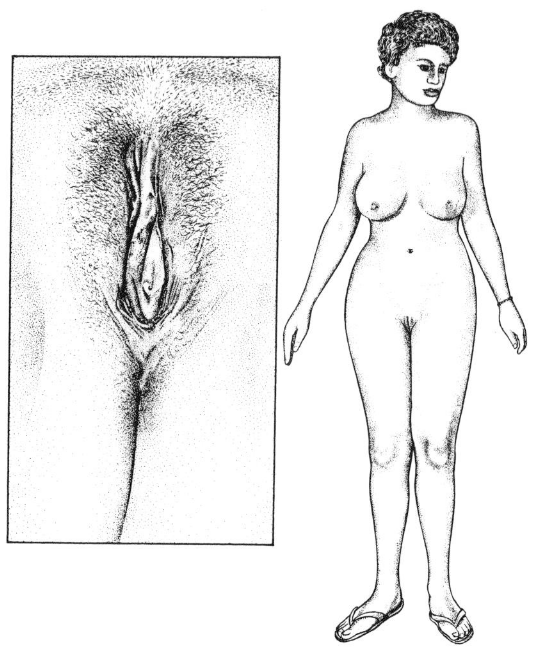

Fig. 5.5. Physical appearance of a genetic male with testicular feminization. Note the normal development of breasts and external genitalia, with scant growth of pubic hair. (From J. Money and A. E. Ehrhardt. *Man and Woman, Boy and Girl*, pp. 95–116. Johns Hopkins University Press; Baltimore (1972).)

Although this is by no means a thorough examination of the human clinical data (see Book 8, Chapters 2 and 3, in the First Edition) it serves to illustrate that although hormones may act on the differentiating brain, the sex of rearing also plays a very significant role in determining subsequent human gender identity. People concerned with animal studies do not consider 'gender identity' or 'psychosexual orientation' but measure patterns of sexual behaviour when animals are experimentally paired. Under these laboratory conditions hormones can be shown to have striking effects on the differentiation of dimorphic patterns of sexual behaviour.

The importance of gonadal hormones for sexual behaviour in adult females

In most female mammals sexual behaviour is clearly dependent on gonadal secretions, and periods of sexual attractiveness and receptivity, or oestrus, appear cyclically, synchronized to the period of ovulation. The main ovarian hormones responsible for this oestrous behaviour are oestrogen and progesterone but the precise mechanism of their action varies among different species. Thus, rodents typically require oestrogen priming followed by progesterone to induce oestrous behaviour, while female carnivores come into heat with oestrogen alone. Progesterone priming is important if oestrogen is to induce sexual behaviour in the ewe, while progesterone secretion may also terminate oestrus and make females refractory to oestrogen in the guinea-pig, rabbit, ferret and ewe. How or why such differences should have evolved in the responsiveness of different species to the two hormones remains a mystery, although oestrogen given alone in sufficient doses will, in time, induce oestrous behaviour in most species.

Actions on the brain. Removal of the ovaries promptly and completely suppresses sexual behaviour in most mammals, monkeys and man being exceptions. Correlations such as these tend to provoke the conclusion that gonadal hormones are activating sexual behaviour by an action on the brain. Certainly, this is the most logical interpretation, but in order to demonstrate conclusively that ovarian hormones affect behaviour by an action on the brain, a number of different approaches had to be adopted. First, it was necessary to show that ovarian hormones reach the brain, and it was not so very long ago that eminent scientists refuted just this point. However, the advent of isotopically labelled steroid hormones soon resolved this dispute, and a number of different brain sites (hypothalamus – medial pre-optic, anterior, ventro-medial and arcuate nuclei – amygdala, septum and midbrain) have been identified by autoradiographic studies as sites of accumulation of the sex steroids (Fig. 5.6). Accumulation of radioactive sex hormones in the brain is not sufficient proof of a physiological action; it is also essential to show that physiological amounts of the hormone can reach this area in sufficient quantities to bind to the receptors.

Brain biochemists have studied the relevant binding affinity of the hypothalamic receptors for gonadal hormones. This can then be related to the availability of the free hormone in CSF, which is in turn related to the non-protein-bound steroid in the plasma, since the transport proteins themselves are unable to cross the blood–brain barrier. Brain biochemists have also made us aware of the need to consider the metabolic fate of steroid hormones after they have become bound to the cytoplasmic receptors (Fig. 5.7). Paradoxically, it is now known that in the rodent brain many behavioural effects of testosterone are dependent on its aromatization to oestradiol. The localization of action of gonadal steroids in the brain, especially within the hypothalamus, has been accomplished by experiments in rabbits, cats, guinea-pigs and rats where minute implants of oestrogen have stimulated sexual behaviour when introduced directly into the brain. These oestrogen implants are too small for the steroid to have peripheral effects on cornification of the vagina, and in fact are active only in restricted areas of the brain, notably the ventro-medial area of the hypothalamus in the female.

Exactly which neurones are affected by oestradiol is not clear and although ultimately influence is brought to bear on the GnRH neurones in the pre-optic area, no one has been able to demonstrate accumulation

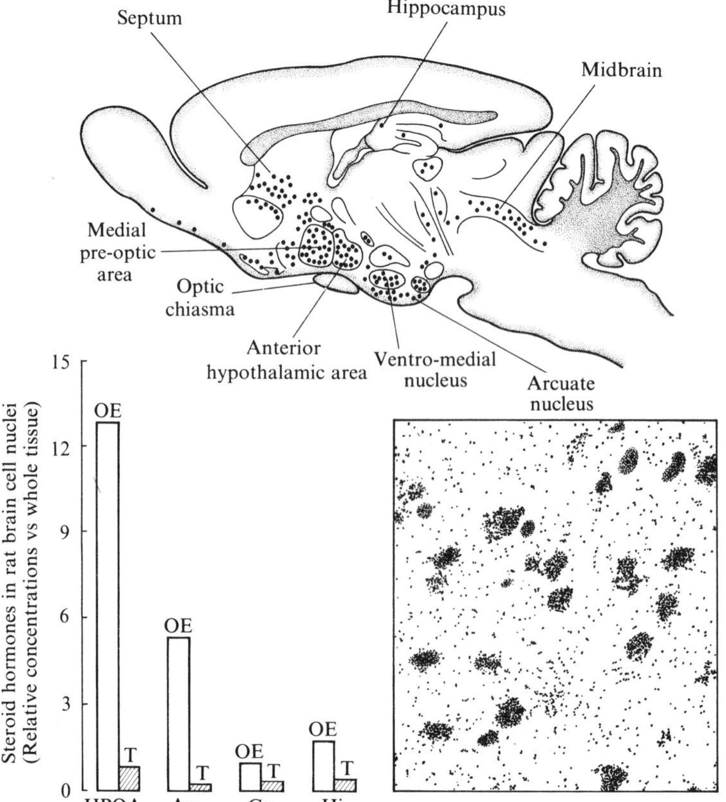

Fig. 5.6. Sagittal section of a typical mammalian brain, with dots indicating the areas of intense accumulation of radioactive steroid after its systemic injection. Insert shows the incorporation of radioactivity (appearing as black silver grains in the radioautograph) in neurones of the medial pre-optic area. The histograms illustrate the relative amounts of oestradiol (OE) or testosterone (T) in the cell nuclei of differing areas of the rat brain following systemic injection of radioactive steroid. HPOA, pre-optic area of hypothalamus; Am, amygdala; Co, cortex; Hi, hippocampus. (From B. S. McEwen. *Neuroendocrinology*, pp. 33–42. Sinnaur Associates; Sunderland, Mass., U.S.A. (1980).)

of oestradiol in those neurones themselves. The essential function of GnRH is to regulate ovarian activity in females, but an increasing number of studies seem to show that injecting GnRH itself either peripherally, or directly into the brain, increases the display of female sexual behaviour (lordosis). Conversely, GnRH antibodies or antagonists injected into the brain, especially in the area of the midbrain periaqueductal grey matter, have been shown to reduce the display of female lordosis behaviour. It is therefore attractive to postulate co-ordinated central and peripheral actions for GnRH. The first would play a role in regulating the onset of sexual behaviour, which in many animals is restricted to the period during which the animal is fertile; the second would ensure fertility by timing ovulation to coincide with the display of sexual activity.

Somatic sites of action. In rodents, the species most thoroughly investigated, oestrogen also facilitates sexual behaviour by increasing the sensitivity and size of the sensory field around the genitalia. Only under conditions of oestrogen treatment does stimulation of the perigenital area produce lordosis, an immobile dorsi-flexion of the rump, enabling the male to mount and achieve intromission. Thus, oestrogen acts peripherally to stimulate sexual behaviour, not only by cornifying the vagina, but also by sensitizing the animal to tactile stimulation (Fig. 5.8). In a number of species, oestrogen also acts peripherally to cause swelling and colouration of the external genitalia (ferrets, dogs and various primate species) which may signal the female's readiness to receive the male. Oestrogen also promotes the male's sexual interest in females by changing the odour properties of urine, or by stimulating an odorous vaginal discharge.

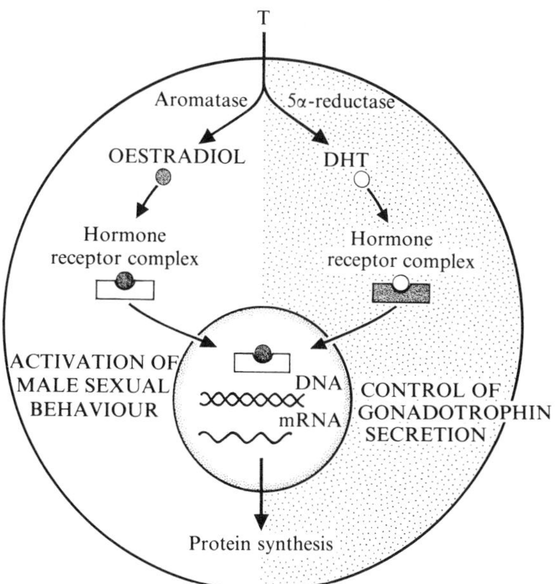

Fig. 5.7. Fate of testosterone (T) on entering the cytoplasm of a neurone, and its subsequent transport to the nucleus prior to transcription and protein synthesis.

Hormones and sexual behaviour in female primates. In adult female primates the effects of hormones on sexual behaviour are not nearly so dramatic as in other mammals, and hence studies have tended to examine in some detail which components of behaviour change. Not only are the different components of sexual behaviour distinguished and measured, but account is taken of the way each animal responds, or fails to respond, to the behaviours shown or olfactory cues emitted by her partner. If that sounds unduly complicated let me illustrate the point with an example.

Female rhesus monkeys in captivity do not have well-defined periods of oestrus but are prepared to accept the male's advances at any time during their cycle. Nevertheless, the sexual interactions of a pair of rhesus monkeys do change significantly during the menstrual cycle, being most intense and with high numbers of ejaculations just prior to the time of ovulation. Somewhat paradoxically the females' sexual invitations to the male may be lowest at this time and increase considerably during the luteal phase of the cycle when ejaculations, the climax of the interaction, have declined to a minimum or stopped altogether. Taken at face value, this would imply that oestrogens make the female less ready to promote sexual interactions, while progesterone enhances her initiative in soliciting behaviour (i.e. her proceptivity). Indeed, injection of these hormones in physiological doses to ovariectomized females would tend to confirm this observation, namely that oestrogen produces little change in female soliciting, while additional progesterone evokes a big increase.

It is only when we obstruct the interaction of the pair by placing a barrier between them, and then make the female work to gain access to the male, that we see a true reflection of the female's inclination to interact with the male. Thus, by providing the female with a lever, which she has to press with some dedication in order to open a door in the barrier separating her from the male, we can obtain a direct measure of her changing interest in that male – we can compare the changes in her rates of lever pressing under differing hormone regimes (Fig. 5.9). With this approach it can be shown that female interest in the male peaks just prior to ovulation and then

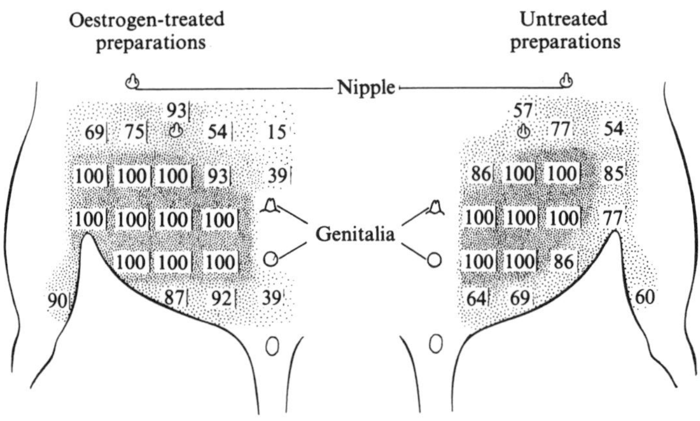

Fig. 5.8. Increase in genital field responses to tactile cues in the oestrogen-treated female rat (left) and in the ovariectomized, untreated female (right). Numbers indicate the percentage of pudendal nerve preparations that produced a burst of activity when tactile stimulation was applied to the area. ('Genitalia' = clitoris and vulva.) (From L. M. Kow and D. W. Pfaff. *Neuroendocrinology*, **13**, 299–313 (1974).)

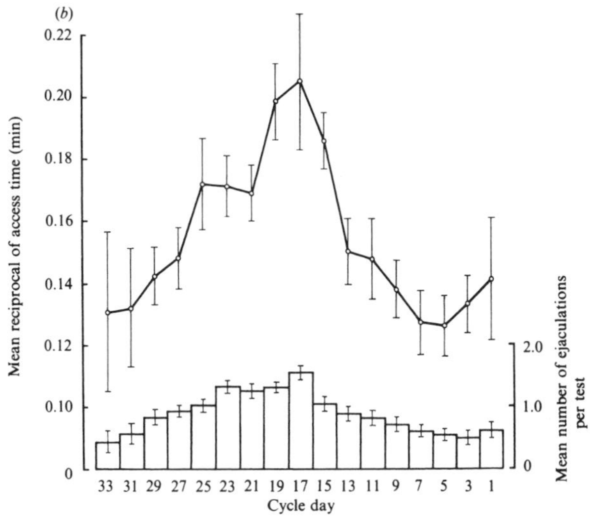

Fig. 5.9. (*a*) Female monkey using an operant lever (top) to open a partition giving her access to the male (middle), and the sexual reward that may follow (bottom). (*b*) The graph shows that the female pressed the operant lever faster and achieved access quicker around the period of ovulation at mid-cycle (averaged over several menstrual cycles). The histograms illustrate that male ejaculation scores also varied with stage of the female's cycle, being maximal at mid-cycle. (Cycle day 1 = first day of menstruation.) However, both female pressing for access and subsequent sexual behaviour occur throughout the cycle. (From E. B. Keverne. *Advances in the Study of Behavior*, **7**, 155–200 (1976).)

declines steeply during the luteal phase of the cycle. This finding is supported in ovariectomized females given oestrogen followed by progesterone, the former hormone promoting faster rates of pressing for access, and progesterone slowing them down. Why then, do these different approaches come up with differing results? The answer seems to be that oestrogen promotes female attractiveness by activating visual and olfactory signals which are emitted at such an intensity that the male mounts without, or before, the female has had an opportunity to solicit with any great frequency. Progesterone decreases the attractiveness of females to males, and increased solicitations therefore represent attempts by the female to maintain the male's sexual interest. Hence, under the influence of progesterone, the female's behavioural signals increase to compensate for the decrease in her visual and olfactory cues that signal female attractiveness.

That this is indeed the case has been shown experimentally by administering these hormones directly into the vagina of the ovariectomized female in minute doses which are insufficient to alter plasma levels significantly and hence reach the brain. Administered in this way, oestrogen stimulates male sexual behaviour without affecting female solicitations, and progesterone inhibits male sexual behaviour while causing a large increase in female sexual solicitations. Since oestrogen increases female attractiveness and thereby immediately provokes male sexual interest, the soliciting behaviour of the female is masked unless the interaction is delayed by some physical separation similar to that in the lever-pressing situation. Similarly, the loss of female attractiveness caused by progesterone decreases the male's interest, resulting in the female partner increasing her soliciting behaviour in an attempt to maintain his interest. Thus we see how a hormone can promote a behaviour not just by acting on the brain, but by decreasing signals in one channel of communication (non-behavioural cues), which is compensated for by increasing signals in another channel of communication (behavioural solicitations).

So far, we have considered the actions of the ovarian hormones, oestrogen and progesterone, and have shown that they act on female attractiveness and proceptive behaviour. (The term proceptive behaviour is used to include all aspects of the female's willingness to initiate a sexual interaction, from the pressing of a lever, full sexual presentation to the male, or even the wink of an eye, if this could be measured with any reliability. Human proceptive behaviour is discussed in Book 8, Chapter 2, First Edition.) We have also mentioned that the female rhesus monkey does not show a circumscribed period of oestrus but will receive the male throughout the menstrual cycle. Does this mean that this primate's brain is, on the whole, emancipated from ovarian endocrine influences, and if so how has this evolutionary change come about?

Certainly a careful examination of laboratory pair tests would suggest there is a large degree of behavioural independence from any distinct neural action of the main ovarian hormones oestrogen and progesterone, these hormones primarily influencing sexual interactions by changing female attractiveness. As outlined above, in rhesus monkeys this is brought about by vaginal cornification, changes in female odour, and the intense redness of the sexual skin (due to vasodilation, which facilitates heat loss and odour dissemination). In some other primate species the ovarian hormones are also important for causing changes in swelling of the sexual skin, which is a physical prerequisite for intromission to occur.

Removal of the main source of female androgens, the adrenals, also affects the receptive behaviour of rhesus monkeys, but in the closely related stumptail macaque this procedure is without effect on female sexual receptiveness. In ovariectomized female rhesus monkeys given oestradiol, removal of adrenal androgens greatly decreases the number of invitations these females make to males, and significantly increases the number of refusals they make to male advances (Fig. 5.10). These unreceptive aspects of sexual behaviour are reversed by low doses of testosterone, given either by injection or by direct implantation into the anterior hypothalamus. We can infer, therefore, that sexual motivation in the female monkey is to some

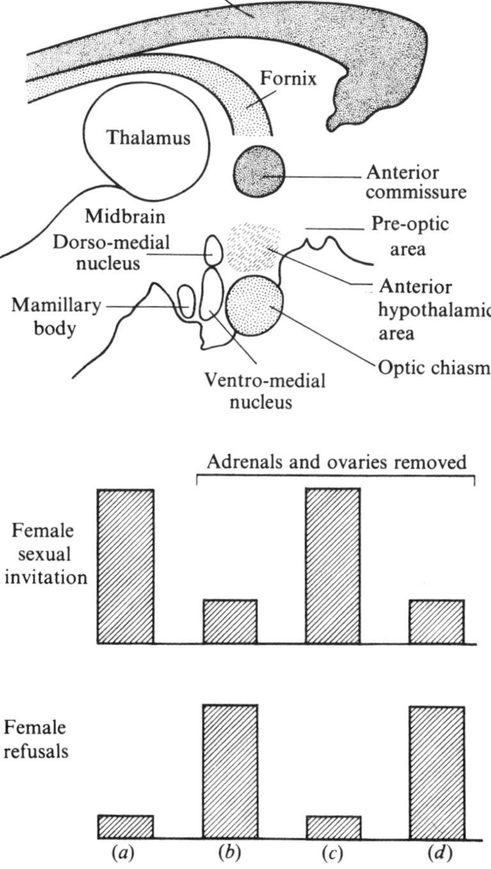

Fig. 5.10. Sagittal section through the diencephalon of a female rhesus monkey brain; implants of androgen in the anterior hypothalamic area reduced female refusals and stimulated sexual invitations. The histograms show changes in these behaviours from those evinced by (*a*) intact animals, to those of ovariectomized and adrenalectomized animals (*b*), (*c*) and (*d*). (*c*) additionally has testosterone implants in the anterior hypothalamus, and (*d*) control implants in the posterior hypothalamus. (From M. Johnson and B. Everitt. *Essential Reproduction*, pp. 186–224. Blackwell Scientific Publications; Oxford (1980).)

extent facilitated by ovarian and adrenal androgens. Whether this is why discrete periods of oestrus are not observed in captivity in rhesus monkeys is an interesting speculation. In terms of the neuroendocrine regulation of sexual behaviour these results suggest that the rhesus brain has achieved a degree of emancipation from the slavish dictates of the gonadal hormones, and this may have been an important concomitant for primate social evolution.

Hormones and sexual behaviour in the male

In mammals, male sexual behaviour is also hormone dependent, as evidenced by the decline in sexual behaviour following castration, and its subsequent reinstatement by testosterone treatment. However, the correlation of sexual behaviour with plasma gonadal steroids is less obvious in the male than the female, some elements of sexual behaviour being retained for weeks, months or even years after castration, long after testosterone and its metabolites have disappeared from the plasma. Moreover, testosterone replacement restores male sexual behaviour only gradually, the dose required and time taken for complete restoration being higher and taking longer, the greater the interval between castration and replacement therapy. It would appear, therefore, that in the absence of testosterone the target tissues become less responsive to the hormone.

Non-primates. In the laboratory rat, again the most thoroughly investigated of all species with regard to male sexual behaviour, sensory spines on the penis disappear following castration. The time course of their degeneration and redevelopment with testosterone replacement closely correlates with sexual behaviour, and hence in this species the peripheral action of testosterone is an important factor in determining male sexual behaviour (Fig. 5.11). It is not the most important factor, however, since dihydrotestosterone (DHT), a metabolite of testosterone that cannot be converted into oestrogen, restores the development of penile spines in the castrate male, but does not restore sexual behaviour unless administered with testosterone or, paradoxically, oestradiol. I say 'paradoxically', but this should come as no surprise when you remember that the action of testosterone on the brain requires the testosterone to be first aromatized to oestradiol. Hence, the effects of testosterone on sexual behaviour in the adult rat can be reproduced by simultaneously giving its peripheral non-aromatizable androgenic metabolite dihydrotestosterone, and its central aromatization product, oestradiol. Testosterone therefore serves as a pre-hormone for the periphery and central nervous system, and the activity of its androgenic and oestrogenic metabolites are essential for the full expression of sexual behaviour.

In red deer, castration of sexually experienced stags, even a few weeks prior to the start of rutting, eliminates all components of sexual behaviour. The animals do not leave their home range to migrate to their traditional rutting territory, nor are they seen to herd hinds or engage in fights with other rutting stags. 'Flehmen' seldom occurs, there is no characteristic rutting odour and the animals do not masturbate. Implantation of castrate stags with either testosterone or oestradiol will restore all components of rutting behaviour in the appropriate season (photoperiod has a major role in triggering sexual behaviour in this species). However, oestradiol, unlike testosterone, is not effective in restoring social aggression if given during the time that testosterone levels are normally falling.

In the red deer stag, the peripheral action of these hormones also has profound effects on behaviour by changing the condition of the antlers. Hormonal manipulations that produce a stunting of antler growth impair both social and sexual success, including sexual aggression, almost as efficiently as castration itself. On the other hand, retention of 'hard horn' antlers gives stags a selective advantage if this occurs when the other stags have cast their antlers and are in velvet.

In most mammalian species some components of male sexual behaviour persist after castration; the components that persist and their frequency of display vary widely not only between species but also among individuals

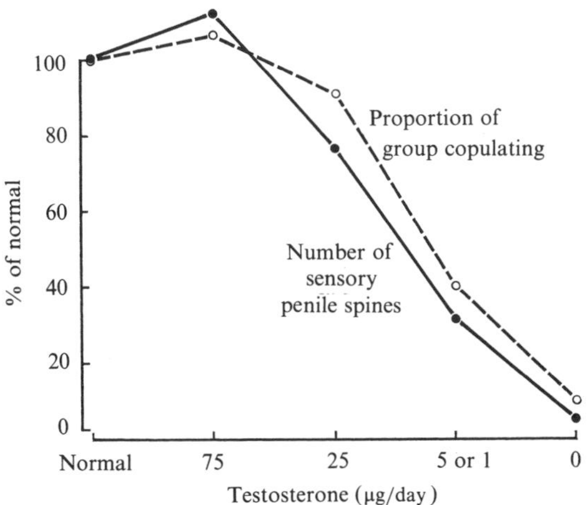

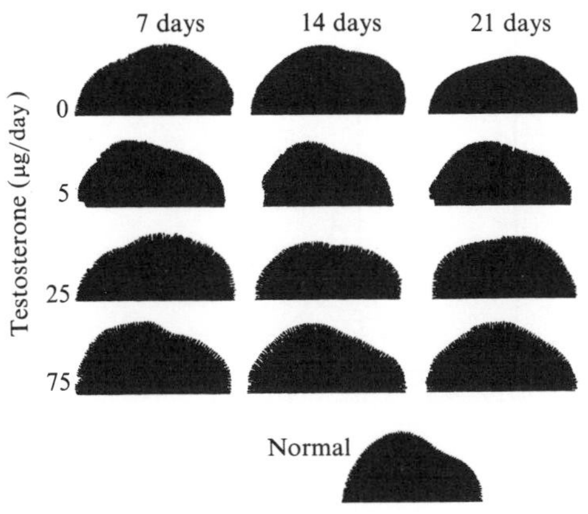

Fig. 5.11. The graph illustrates the correlation between the number of penile spines and the sexual behaviour in a population of castrated male rats given decreasing doses of testosterone. 'Normal' refers to the intact animal. Below are cross-sections of the penis, showing the development of penile spines in castrated males given increasing doses of testosterone (From F. A. Beach and G. Levinson. *J. Exp. Zool.* **114**, 159–71 (1950).)

within a species. In general, the ejaculatory response disappears first and is followed by the disappearance of intromission and mounting. Although rodents lose their ejaculatory response within a few weeks, this continues much longer in cats and may persist for some years in dogs and rams and monkeys.

Primates. Studies of sexual behaviour in male primates have languished somewhat, perhaps because of the investment in time necessary to find any effects of castration. Sexual behaviour tends to persist in adult castrated male rhesus monkeys for at least a year after castration in the laboratory, while castrate males released into free-ranging groups of rhesus monkeys have been observed mounting as much as 7 years later. As in female primates, if the components of sexual behaviour are examined carefully in the laboratory, some effects of castration appear within as little as 5 weeks. The first problem a castrate male encounters is in achieving intromission, which greatly reduces the number of ejaculations achieved in the limited time of an experimental testing period (Fig. 5.12). Nevertheless, male rhesus monkeys will continue to mount females for years, and when they do eventually stop it is difficult to determine if their lack of motivation is in any way directly related to the absence of testosterone or

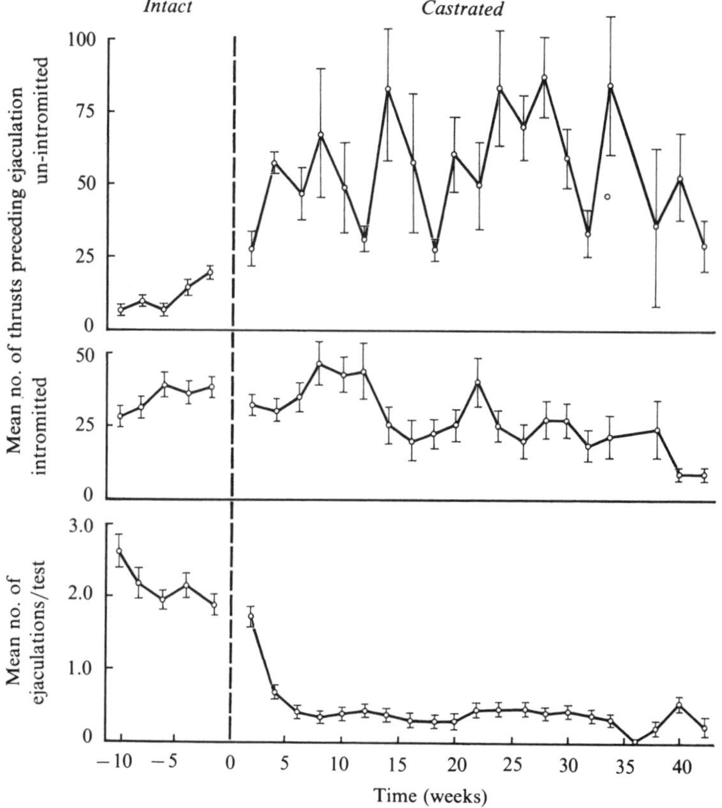

Fig. 5.12. Effects of castration on various components of sexual behaviour in male rhesus monkeys. Ejaculation scores decline rapidly within 5 weeks of castration, but the male's interest in the female is maintained and some elements of sexual behaviour (un-intromitted thrusting) increase. (From data published in R. P. Michael and M. I. Wilson. *Endocrinology*, **95**, 150–9 (1973).)

to the adverse learning experience during their impaired sexual performance. Some female partners may become aggressive towards castrate males, especially when the males repeatedly mount without intromission, and therefore other behaviours may need to be taken into account when assessing the sexual relationship. The length of time sexual interactions continue after castration in primates depends largely on the female partner and the amount of past sexual experience, rather than on gonadal hormones. Indeed, a male monkey can be induced to continue sexual behaviour for longer periods after castration by simply changing the female partner. On the other hand, he will stop much sooner if other males are introduced that compete for the female partner.

The decline in sexual behaviour of male monkeys following castration has much in common with the decline in their sexual behaviour following section of the dorsal nerve of the penis. The dorsal nerve conveys sensory information from the specialized spines of the glans penis, spines which atrophy following castration. This would suggest that the key site of action for testosterone in determining the sexual behaviour of male primates is peripheral rather than central. This suggestion finds support in experiments that have demonstrated that dihydrotestosterone alone, the peripheral metabolite of testosterone, is sufficient to restore sexual behaviour to normal in male monkeys castrated some 3 years prior to treatment.

The experiments that I have just described have enabled scientists to look at the mechanisms and sites of hormone action in testing situations that isolate male monkeys from their normal social environment. It is quite clear from these studies that sexual motivation is little affected by removal of gonadal androgens, and any long-term changes in sexual interaction can be accounted for by examining the somatic sites of hormonal activity. One cannot, however, predict from these studies that castration of a group of free-ranging males would likewise have little effect on sexual behaviour, even if the peripheral metabolite dihydrotestosterone is administered. Testosterone may have a central action affecting assertive or aggressive behaviour, which in turn may influence a male's social rank and thus his competition for and accessibility to females (this point is taken up again later).

The extent to which pair-test studies on the rhesus monkey can be extrapolated to the human male is somewhat questionable. There are no penile spines in man, and no evidence that sensory feedback from the penis in sexual arousal is dependent on testosterone or its metabolites. In the adult human male ejaculations will cease within weeks of castration but 'dry-run orgasms' continue with corresponding feelings for months and even years after castration. The length of time before erectile potency is lost is also extremely variable, but does appear to be under the influence of testosterone. Hence we are left facing the same problem of interpretation in man as for the monkey, namely the extent to which sexual motivation is weakened directly through the loss of androgens, or indirectly through the loss of performance. This is an important consideration for the treatment of sexual dysfunction in man, and especially for the treatment of sex offenders whose sexual motivation is in need of restraint. In the latter context, ethinyl oestradiol, an oestrogen, or cyproterone acetate, an anti-androgen, have been reported to suppress sexual thoughts and masturbation without significantly affecting erectile potency. This would suggest that in man androgens may indeed enhance sexual arousability by acting on the brain.

Sensory systems and sexual behaviour

Sensory input is extremely important in regulating the sexual behaviour of mammalian species. Monitoring changes in the sensory environment is a function of the special senses, and in many cases a special part of these senses that is unrelated to perceptual awareness of the environment (retinohypothalamic projection; accessory olfactory amygdaloid projection). Which particular aspect of the environment is important varies according to the species, with certain species responding more to olfactory and others to tactile and other sensory cues for sexual arousal.

Light is a very important regulator of mating activity (see Chapter 4). Most rodent species display sexual behaviour only during the hours of darkness. Moreover, changes in day-length are important for initiating sexual behaviour in seasonally breeding mammals. In general, species such as ferrets and field mice which mate in the spring can be induced to mate sooner by artificially increasing day-length, whereas autumn maters, such as sheep, goats and red deer, require a reduction of day-length to enhance the early onset of mating activity. Of course, this onset of sexual activity in seasonally breeding mammals is secondary to light-induced changes in gonadal activity.

However, it has been shown in rats that despite the absence of any annual rhythm in reproduction, there is a well-defined circadian rhythm in neuroendocrine and motor activities. Circadian changes also occur in the sexual behaviour of the female rat, and these are not secondary to secretory activity of the gonads. Thus, female sexual behaviour (lordosis) declines during the day despite the continued presence of oestradiol and progesterone administered to ovariectomized females (Fig. 5.13). Moreover, ovariectomized females given oestradiol alone also show high levels of lordosis behaviour in the dark portion of the light–darkness cycle and low levels during the daylight. This rhythm is abolished by lesions to the suprachiasmatic nucleus, the area of the hypothalamus that receives the direct retinal projection, suggesting a direct neural control by photoperiodic stimuli on female sexual behaviour.

Tactile stimuli play an important part in determining sexual co-operation and co-ordination in many species. The male rat experiences great sexual difficulties when the female is unco-operative or even when she adopts an abnormal standing position. Intromission is accomplished only when the female dorsi-flexes her back (lordosis), directing the vagina into an accessible position. The male rat evokes this response by a spatially and temporally ordered sequence of cutaneous stimulations of the female's lumbar and perivaginal region. It is not surprising then, that desensitization of these areas on the female's body (by anaesthesia or denervation) reduces lordosis behaviour and consequently the number of successful mounts by the male. Lordosis can, however, be elicited even in decorticate females, and so is perhaps better considered as a diencephalic reflex which does not require a cognitive awareness of the tactile input.

Tactile stimulation during coitus, or as a result of manual vaginal probing by a human investigator, facilitates sexual behaviour in the female rat, but it terminates the sexual co-operation of female guinea-pigs. Without the experience of coitus, lordosis can be elicited repeatedly for up to 8 h. Following tactile stimulation of the vagina, lordosis can be elicited for only 1–2 h. This tactile effect is not dependent upon the presence of the ovary, but appears to be a direct consequence of tactile signals on neural mechanisms involved in sexual behaviour.

De-afferentation of the penis in rats and monkeys produces major deficits in sexual behaviour, particularly the ejaculatory response. The mounting behaviour of rats becomes disorientated, they fail to intromit and ejaculate, and eventually they abandon any attempts at sexual interactions. In monkeys, changing the female partner will often reinvigorate such nerve-sectioned males so that they start mounting again. Although these effects are independent of any hormonal influences, they are in many respects very similar. Thus, castration of male rhesus monkeys also produces a great reduction in intromission, without which ejaculation likewise fails to occur. The reason for these similarities is probably because

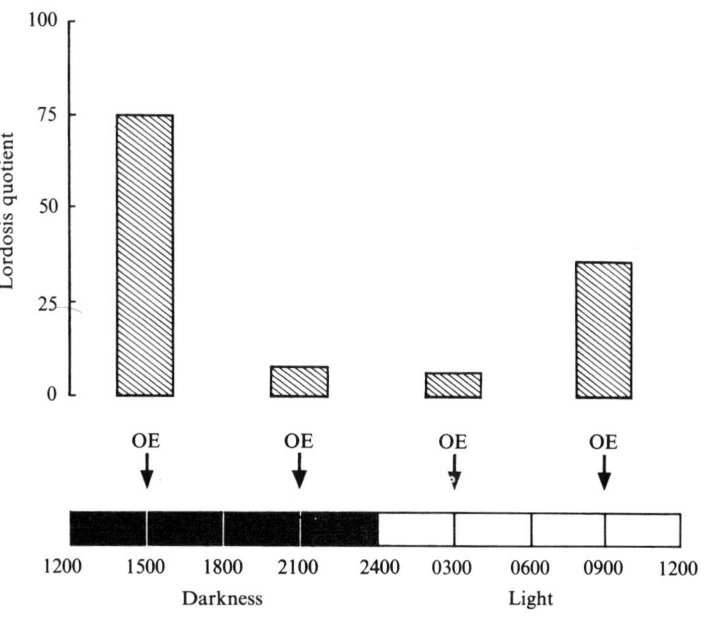

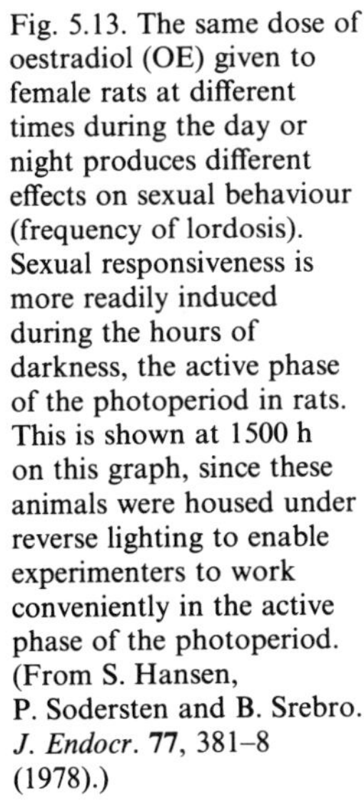
Fig. 5.13. The same dose of oestradiol (OE) given to female rats at different times during the day or night produces different effects on sexual behaviour (frequency of lordosis). Sexual responsiveness is more readily induced during the hours of darkness, the active phase of the photoperiod in rats. This is shown at 1500 h on this graph, since these animals were housed under reverse lighting to enable experimenters to work conveniently in the active phase of the photoperiod. (From S. Hansen, P. Sodersten and B. Srebro. *J. Endocr.* **77**, 381–8 (1978).)

sensory feedback from the glans penis is necessary for correct orientation which in turn is prerequisite for intromission. While removal of gonadal hormones results in atrophy of the penile spines, dorsal nerve section effectively denervates the glans penis.

Olfactory cues may also regulate sexual behaviour, usually through the mutual attraction between male and female, but also by bringing animals into a state of sexual readiness. Thus, large groups of female mice living together enter a state of semi-permanent dioestrus. Introducing males, or their odour, will promote the onset of oestrus 3 days later by reinstating normal ovarian activity. Sexual behaviour is in this case secondary to odour-induced changes in neuroendocrine responsiveness. Olfactory cues can, however, promote sexual behaviour more directly. In pigs, odours from the boar produce an immobilization reflex in the oestrous sow which enables him to mount her (see Book 3, Chapter 6, Second Edition). Commercial use has been made of this discovery by developing aerosol sprays containing boar odours (Δ^{16} steroids) which can be used to detect oestrus in isolated sows to facilitate artificial insemination. In other species (mice, hamsters), odours from the female sexually arouse males, causing increases in plasma testosterone (Fig. 5.14), while in the ewe, odours from the ram will induce immediate pulsatile discharges of LH, sometimes resulting in premature resumption of oestrus and ovulation (see Book 3, Chapter 1, Second Edition). In the hamster, vaginal secretions from oestrous females provide sufficient sexual stimulation to induce males to attempt to mate with other males that have these secretions applied to their fur. Even an anaesthetized male hamster can be transformed into a sexually appealing subject for other males if treated in this way.

Olfactory cues, where they have been shown to influence mammalian sexual behaviour, have often been called pheromones. One complication of using the term 'pheromone' in the repertoire of mammalian behaviour

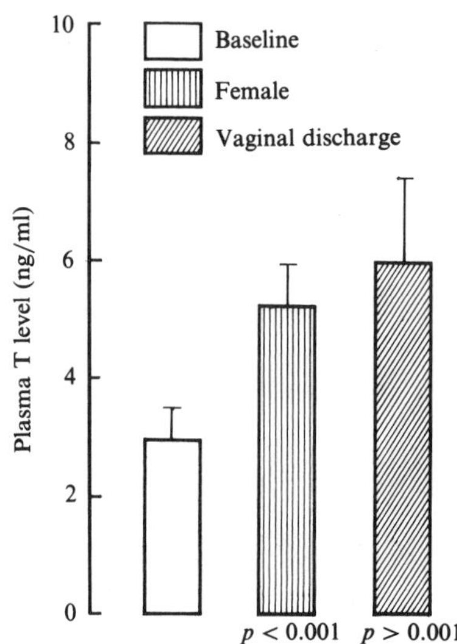

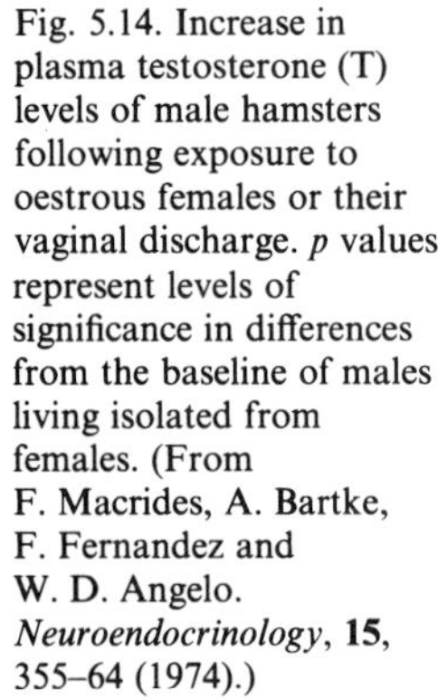
Fig. 5.14. Increase in plasma testosterone (T) levels of male hamsters following exposure to oestrous females or their vaginal discharge. *p* values represent levels of significance in differences from the baseline of males living isolated from females. (From F. Macrides, A. Bartke, F. Fernandez and W. D. Angelo. *Neuroendocrinology*, **15**, 355–64 (1974).)

stems from the two types of evoked response which are not necessarily mutually exclusive. If the pheromone produces a more-or-less immediate change in the behaviour of the recipient, it is said to have a releaser effect, and sex attractants constitute a large and important category of the releaser pheromones. Primer pheromones, on the other hand, alter some feature of the animal's physiology, and this may, in turn, have consequences for behaviour. However, the latter type of response is slow to develop, and in mice it demands a prolonged stimulation, which is mediated through the vomeronasal, accessory olfactory system and the endocrine system. Thus, olfactory cues can influence sexual behaviour in at least two ways, one purely neural and the other involving the pituitary and gonads. Removal of the olfactory bulb from an animal may, therefore, decrease the likelihood of its finding an attractive mate, but the operation can also result in underdeveloped gonads, influence ovulation, oestrous cycle and implantation, and impair sexual performance. Thus, when we consider olfactory cues and mammalian sexual behaviour it is important to know which response we are observing, and while this is fairly evident when the animal is presented with specific odour cues, it is not so clear when the experimental procedure involves olfactory deprivation, especially by olfactory bulb ablations, which destroys both the main and accessory olfactory systems.

Although the sensory modalities that are relied upon to facilitate or modulate behaviour differ among mammalian species, in no species is any one of these channels of communication essential for sexual behaviour. In the natural sequence of mating behaviour most mammals depend on multiple sensory cues to assess their environment. Studies involving surgical lesioning of different sensory receptors in a number of species have demonstrated that ablation of any single sensory modality does not prevent copulation. Primates, perhaps more than any other mammal, exemplify this independence of sexual behaviour from specific sensory signals. Neither tactile, visual nor olfactory cues determine their behaviour in a stereotyped manner. It is therefore interesting to find that in these highly evolved social mammals, the social environment has most influence on the display of sexual behaviour.

The social environment

The social environment is known to influence many aspects of reproductive performance in several mammalian species, either by a direct action on the expression of behaviour or indirectly by modifying hormonal state. Thus, when male mice, hamsters, rats, rabbits, the ram and the bull are presented with a sexually stimulating oestrous female, their testosterone levels are likely to rise. On the other hand, social stress and overcrowding can suppress gonadotrophin secretion and reproductive behaviour. However, it is only when we consider group-living primates that the significance of the social environment in relationship to sexual behaviour becomes all important, not only for the ways in which it restricts sexual behaviour, but also for the way it enriches its development.

In contrast to other mammals that are social during only part of the year, most primates live throughout the year, and indeed throughout life, in social groups. For an individual monkey, its life is the social group, in all its richness, its complexity and its stability. Life outside the social group vastly increases the likelihood of death from predation, and any mobility that does occur outside the social group invariably involves movement to another troop. Even from the very early stages of life, when the infant monkey knows and is concerned with only a very small part of its group, what happens in the group nevertheless reflects on that young member as well, both directly and indirectly through its mother and peers. These early social experiences, especially with peers, are essential to the development of normal adult sexual behaviour. Harlow has shown that monkeys raised with peers, but without a mother, grow up to be normal adults, whereas monkeys raised with only a mother show little play and no sexual

Fig. 5.15. (*a*) Baby monkey reared in social isolation, huddled up and sucking its fingers. A monkey of the same age and given a cloth surrogate mother (*b*) shows less huddling and is more advanced in behavioural development. (From H. F. Harlow and M. K. Harlow. *Sci. Amer.* **207**, 136–46 (1962).)

(*a*)

(*b*)

behaviour when adult. Infants deprived of both maternal and peer interaction grow up to be very abnormal indeed (Fig. 5.15). Not only is mating behaviour absent, but all social interactions are reduced and even in the presence of other normal monkeys, these socially deprived animals fail to interact and often inflict injury upon themselves. Sexual behaviour of primates develops, therefore, in a very complex manner, involving not only the hormonal milieu, but the social environment as a whole.

Somewhat restricted social environments, such as peer groups that are all male or all female, also operate to determine patterns of sexual interactions. Rearing infants in mother–infant groups where play contact was permitted with peers of both sexes resulted in clear differences in 'foot-clasp mounting' frequencies between males and females. Males mounted a great deal, and females very little. But when peer contact was restricted to the same sex, females increased foot-clasp mounting and males showed very little of this behaviour (Fig. 5.16).

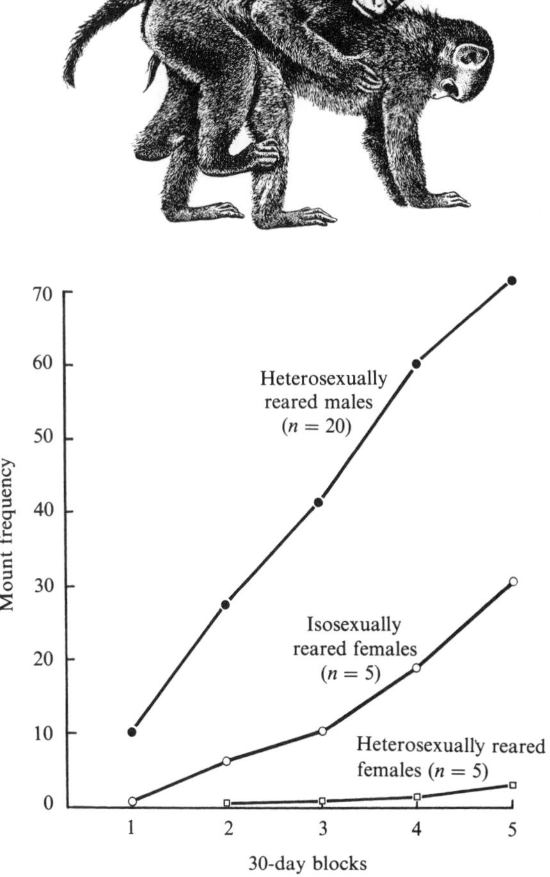

Fig. 5.16. Development of the foot-clasp mount shown here is more rapid and greater in heterosexually reared males than it is in heterosexually reared females. Females reared in isosexual groups, however, show an enhancement of this behaviour. (From D. Goldfoot and K. Wallen. In *Recent Advances in Primatology*, pp. 155–60. Academic Press; New York (1978).)

Dominance and sexual behaviour

Dominance is a concept perhaps best defined in terms of the outcome of aggressive interactions. Thus, if animal *A* hits *B* and *B* fails to retaliate or initiate aggression towards *A*, then *A* is said to be dominant to *B*. It does not necessarily transpire from this definition that animal *A* is the most aggressive in his group. In other words it is the direction and not the amount of aggression that defines the hierarchy. In well-established social groups of monkeys, actual fighting may be extremely rare, and the dominant male is often able to settle conflicts simply by a look or stare at the more subordinate monkey. Physiologists tend to use the term 'dominance' as if it imparted some definable feature to an individual, or as if it had some quality of its own. If this is indeed true, no one has so far discovered what the property might be, and there is no reliable way of predicting a dominance hierarchy on the basis of any measurable parameter when examining individual monkeys in social isolation.

In many free-living primate species, sexual behaviour is also correlated with rankings in the dominance hierarchy of their social group. Dominant male rhesus monkeys in India spend more time in consort with females and do more mounting than their subordinates. This may be related to a priority for access, dominant males gaining the first access to sexually attractive females. If only one female in a group is attractive at a given time then only the first-ranking male will mate with her, but when two females are simultaneously attractive, the first- and second-ranking males will mate with them, and so on.

The notion of social dominance has not been without its critics, especially when considering captive studies where experimenters have imposed hierarchial behaviour on monkeys by restricting space, and making them compete for food or avoid electric shocks. Nonetheless, hierarchies based on dominance in agonistic encounters in the wild have proved to be a useful descriptive tool, and in many primate societies have often proved to be predictive of male mating success. There are reports of strong, positive correlations of dominance rank and reproductive behaviour for macaques, baboons, vervets, langurs and patas monkeys. Such field observations are not necessarily indicative of breeding success, unless some other indicator is also available that enables the observer to determine whether or not the female is at the fertile stage of her cycle. Nevertheless, in the social group, sexual behaviour is not equally distributed among males, and those of higher rank get more than their share, while the most subordinate males may be totally excluded (Fig. 5.17).

Since the first section of this chapter dealt in some detail with the way in which gonadal hormones determine sexual behaviour, it is logical to enquire whether gonadal hormones are themselves influenced in any way by the social hierarchy. Could it be, for example, that the lower levels of sexual behaviour seen in subordinate males are related to low levels of plasma testosterone? A number of studies have indeed shown that male rhesus monkeys of high rank have higher testosterone than subordinates, and a few scientists have been tempted to relate these findings causally to the differing levels of sexual behaviour. However, these conclusions must be considered erroneous for a number of reasons. As I pointed out earlier, total castration of male rhesus monkeys is without any significant effect on sexual behaviour in the short term. Secondly, in a social group of long-term castrate talapoin monkeys, testosterone administration restores sexual behaviour in only high-ranking and not low-ranking castrates. Finally, testosterone implants given to low-ranking intact males do not enhance their sexual behaviour or, for that matter, alter their position in the social hierarchy. Therefore low levels of testosterone, although

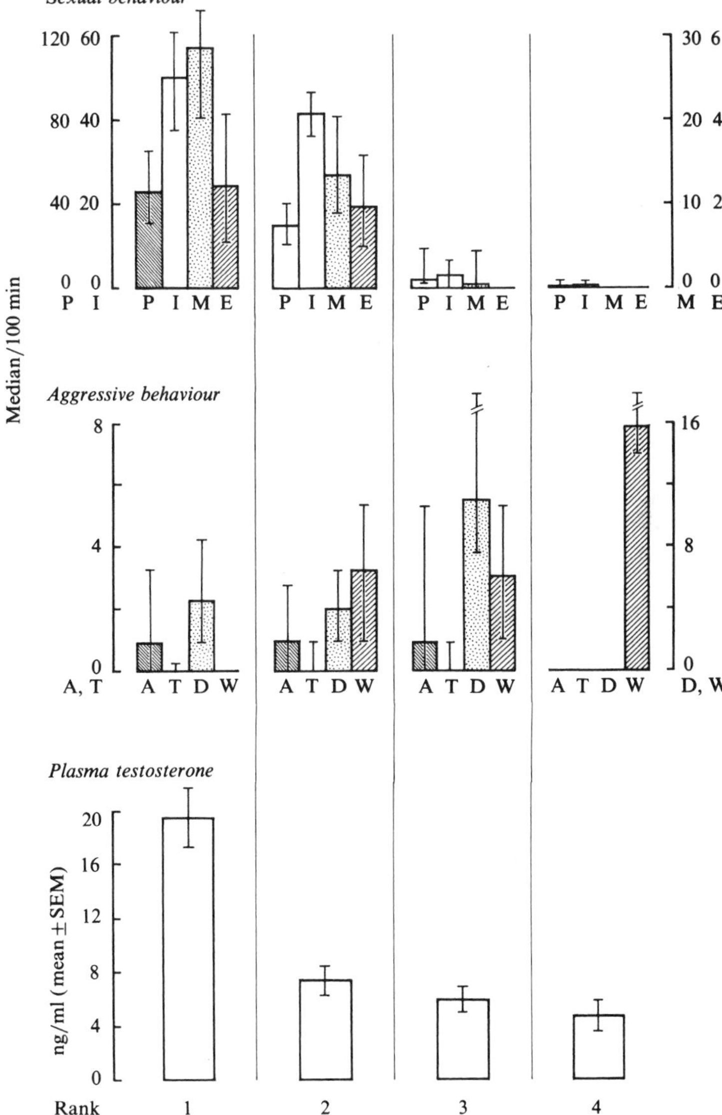

Fig. 5.17. Effects of the social hierarchy on sexual and aggressive behaviour of male talapoin monkeys, and on their plasma testosterone levels. The highest-ranking male shows more sexual behaviour, receives least aggression and has highest levels of plasma testosterone. Sexual behaviour: P, presentations received; I, inspections of female genitalia; M, mounts; E, ejaculations. Aggressive behaviour: A, attack received; T, threat received; D, displacement made (mild aggressive act); W, withdrawal from aggression. (From E. B. Keverne. In *Sex, Hormones and Behaviour*, pp. 271–98. Ciba Foundation Symposium 62 (new series). Excerpta Medica; Oxford (1979).)

correlated with social subordination, are neither a determinant of social rank nor causally related to the depressed levels of sexual behaviour seen in these males. More likely, the sexual behaviour of higher ranking males is responsible for their elevated testosterone, i.e. it is because dominant males show sexual behaviour that they have high testosterone, and not the converse. Likewise, social subordinates have low plasma levels of testosterone because they do not show sexual behaviour and live permanently anxious of impending aggression from more dominant males. It is this continual threat of aggression that inhibits their sexual proclivities and not the reduced levels of testosterone. In other words, subordinate monkeys have learnt that if they come into conflict with the dominant male they are likely to receive a thrashing, and as sure a way as any to come into conflict with dominant males is to compete for females. Hence, whenever low-ranking males are occasionally seen consorting with females, it is usually out of sight and sound of the dominant male.

If elevated testosterone in the dominant male is not causally initiating his sexual behaviour, then what it is doing? Is it just a physiological side-effect resulting from increased hypothalamo-pituitary activity with no consequences for behaviour? So far, no behavioural significance has been clearly attributed to these higher levels of testosterone, although physiologically they are integral to sustained spermatogenesis. Suggestions have been put forward linking high testosterone with increased male attractiveness, and it is true that dominant males receive considerably more attention and soliciting from females than do subordinates. However, this argument becomes tautological, since the females may be soliciting dominant males because these males are more sexually active.

If, as I have suggested, one of the constraints on sexual behaviour of subordinate males is the impending threat of aggression from higher ranking males, then it might be predicted that removal of higher ranking males would permit the subordinate male to display sexual behaviour. This prediction has been put to the test in social groups of talapoin monkeys and the outcome was somewhat surprising. Subordinate males, each given access to their females in the absence of dominant males, showed very little sexual behaviour, and even this infrequent sexual behaviour was of a low performance quality. Although plasma testosterone was seen to rise in these males, so too were plasma cortisol and prolactin, the so-called 'stress' hormones. It would appear, therefore, that one of the correlates of chronic social subordination is an inability to cope with sexual behaviour even when the dominant male is removed. Whether this inability is a consequence or determinant of social subordination is currently being investigated.

Dominance hierarchies as assessed from the direction of aggressive encounters have also been described in females of a number of primate species (macaques, baboons, talapoins, gelada baboons). Aggressive behaviour is less frequent among females than among males, often making the determination of their hierarchy a difficult task from observations in their natural habitat. A female's position in the social hierarchy can nevertheless have important consequences for other behaviours, especially sexual behaviour. High-ranking females seem to attract more sexual attention than lower-ranking females, although low ranking females are not totally excluded from sexual interactions (Fig. 5.18). Again we are faced with the question whether this unequal distribution of sexual behaviour can be accounted for in terms of endocrine differences that determine the

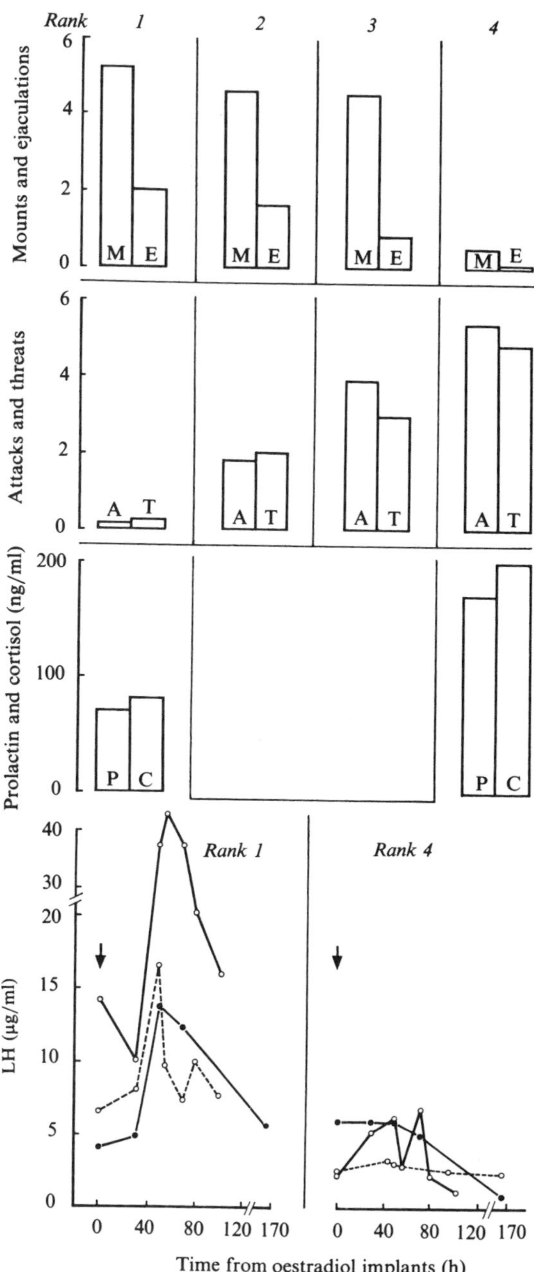

Fig. 5.18. Distribution of sexual and aggressive behaviour according to rank in female talapoin monkeys (ovariectomized and given the same dose of oestradiol). M, mounts; E, ejaculations received; A, attacks received; T, threats received. In the lower section are shown the levels of stress hormones prolactin (P) and cortisol (C) in highest- and lowest-ranking females, and the LH response to the challenge of positive oestrogen feedback (arrow). No LH surge occurs in the female of rank 4. (From L. Bowman, S. Dilley and E. B. Keverne. *Nature*, **275**, 56–8 (1978).)

attractiveness of females. Is it possible for the harassment associated with low rank to suppress ovarian function? Field studies of gelada baboons and captive studies with marmosets would suggest that this could be a distinct possibility, since the subordinate females of the former have longer and infertile cycles, and marmosets of low rank fail to ovulate. However, an endocrine basis for differential attractiveness does not in itself account for the uneven distribution of sexual behaviour, since such differences still persist in social groups of talapoin monkeys where the females are all ovariectomized and receive the same amount of oestrogen, the ovarian hormone responsible for enhancing attractiveness. Moreover, even selective oestrogen treatment of only the lowest-ranking female does not enhance the sexual attention she receives from males, although such treatment considerably enhances the aggression she receives from higher-ranking females. It is this aggression, or the potential for aggression, that restricts her sexual interactions. Dominant females from a variety of primate species (marmosets, talapoins, macaques, baboons) have been observed to aggress and interfere with the mating of subordinates. In talapoin monkeys, such harassment produces in the recipient an elevated cortisol and prolactin level, and a failure to show ovulatory LH surges when challenged with positive oestrogen feedback (Fig. 5.18). Hence, in females just as in males we see that behavioural interactions can have consequences for the reproductive hormones. While the ovarian hormones are a necessary prerequisite for sexual interaction, their role becomes a permissive one, and the social environment determines who is to mate. Moreover, in those monkeys that are not permitted sexual interaction we see that reproductive activity may itself go into decline. In many respects the *status quo* of the group is thereby maintained, since inhibition of ovarian activity will also reduce the attractiveness of low-ranking females, which in turn will reduce the need for subsequent overt aggression.

In conclusion, then, we see that gonadal hormones play a very significant role in determining the reproductive behaviour of a number of mammalian species. The production of these hormones, and hence the behaviour, is often under the influence of the physical environment, especially in seasonal breeders. In rodent species which breed throughout the year, discrete periods of oestrus nevertheless persist, and although not seasonally determined, these periods of oestrus are still regulated by diurnal changes in the physical environment. Thus, whatever holds the key to gonadal activity also regulates, albeit indirectly, sexual behaviour.

The sexual behaviour of monkeys, apes and man is not, however, so strictly regulated by gonadal hormones. Studies of several monkey species in captivity (*Macaca mulatta* the rhesus monkey, *M. fascicularis* the crab-eating monkey, *M. nemestrina* the pig-tailed monkey, *M. arctoides* the stump-tailed monkey, and *Miopithecus talapoin* the talapoin monkey) have shown females to be receptive at all times during their cycle, and even in apes like the chimpanzee with a large sexual skin swelling, mating has been observed throughout 30 per cent of the menstrual cycle. Likewise in the gorilla, sexual behaviour is clearly not restricted just to the mid-cycle period but may occur for up to 14 days of the cycle, while captive pairs of orang utans have been observed to mate on every day of the menstrual cycle. It is true that in their natural environment some of these primate species are seasonal breeders, and all of them show what field workers have described as discrete periods of oestrous behaviour. When such primates are removed from their native social groups and brought into captivity, their sexual behaviour may continue throughout the year and at all times during the females' menstrual cycle. In other words, the physiological potential is available for males to be continuously sexually active and females to be almost permanently sexually receptive. Such extended sexual activity does not prevail in non-primates or lower primates even in the laboratory.

Of course, these monkeys and apes have not evolved in a laboratory and the fact that reproductive behaviour in their natural environments is restricted to discrete times in their cycles and in the year must not be overlooked. Solly Zuckerman in his book *The Social Life of Monkeys and Apes* made the mistake of basing his theory for the social evolution of all primates on the study of a single species, mainly in a zoo. Zuckerman observed an 'uninterrupted sexual and reproductive life' and there was 'no implication that the sexual stimulus holding individuals together was ever totally absent'. For the next 25 years this theory dominated thought in primate sociobiology. Given the evidence available at the time, Zuckerman's conclusions were not unreasonable, but subsequent field studies have shown that a large variety of temperate-zone primates, all with cohesive societies, breed with a seasonal periodicity. Moreover, many of the important correlates of social structure are non-sexual and involve defence against predators, territoriality and competition for food resources. That is not to imply that sexual strategies play no part in determining social organization, but they are clearly not the great unitary explanation of primate social evolution. Nevertheless, non-human primates are unique among mammals in that much of the sexual behaviour observed in their natural habitat is not reproductive, and even the most experienced of field workers are led to conclude that non-reproductive sexuality functions to maintain troop cohesion and reinforce social bonds.

Obviously, sexual behaviour does not have to be continuous to be always present. Any psychologist will testify to the fact that intermittent rewarding behaviour is more difficult to forget (extinguish) than continuously reinforcing behaviour. Hence, the absence of overt sexuality does not exclude it from what Robert Hinde has called the 'deep structure' of society. With the development of a larger brain and presumably a greater capacity for memory, so the possibility arises that the sight, sound and smell of conspecifics may become rewarding as a consequence of these cues being coupled with previous hedonistic experiences. How else can we explain the partner preferences that are sufficiently robust to override the temptation of attractive females when the preferred partner is in a state of anoestrus? Of course, there are many aspects of social interaction other than copulation which reinforce primate social bonds. The touch of conspecifics may be particularly important, especially during intense reciprocal grooming periods.

No one has yet been able to determine what the proximal factors are that regulate primate seasonal breeding in the wild, and in those equatorial monkeys that are seasonal it is extremely difficult to imagine which aspects of the natural environment they are responding to, and why. Even when brought into captivity, rhesus monkeys continue to display annual rhythms of sexual activity which are in synchrony with their behaviour in the wild. The major difference between field and laboratory is the amount of time that is spent in sexual interactions. It has been suggested by Butler that 'the continuous erotic behaviour seen in captive primates must be regarded as a deviation resulting from the boredom of captivity'. But, more importantly in captivity they would find themselves emancipated from the restrictions imposed by surviving in their natural habitat (widely dispersed food resources, predators, competitors, etc.) and therefore able to express their physiological potential more fully.

In the natural habitat, the limiting resource that inhibits the further increase in population size of most species is usually food availability. When a species has exploited to the full the carrying capacity of its environment, then food availability represents a major factor in evolution. Intraspecific competition for resources will, at this stage, become intense and selection will favour efficiency of resource utilization combined with a relatively low rate of reproduction. Here there will be selection for efficient production of offspring and for adaptations that enhance the survival of those offspring in a highly competitive environment (*K* selection) (see the following section on maternal behaviour). Primates as a group are typically *K*-selected with respect to many other mammalian groups and usually give birth to only a single offspring. One of the characteristic features of *K*-selected primates is that the young develop slowly and so have a long period of maternal dependency. The most important determinant of lifetime fertility in these species is the length of the interbirth interval, especially in equatorial primates where photoperiod is not of overwhelming importance. Roger Short has drawn attention to the importance of lactation for inhibition of ovulation, which may range from 420 days in the olive baboon and 2.5 years in the gorilla to 3.5 years in the chimpanzee. This has expanded the time during which the mother can provide her offspring with undivided attention and transmit the benefits of her social experience. During this time there is rapid brain growth, which has provided not only the substrate for socialization, but a large cortex endowing primates with the potential to adapt their mating strategies according to a new or changing environment.

This not only applies to the changed conditions met on moving into the laboratory, but was demonstrated in the field by Hans Kummer who transferred the socially polygynous female anubis baboon to a restricted harem society of the hamadryas baboon, and vice versa. Females of both species rapidly adapted behavioural strategies to suit their new social environment. Likewise on the island of Siberut (Malaysia) the Mentawai snub-nose langur lives in both polygamous and monogamous groupings. In the southern part of the island they have adopted a monogamous strategy, whereas those groups living in the more densely populated north are polygynous. It is because primates are not 'hard-wired' for their sexual behaviour that they have this capacity to adapt their social strategies to a changing environment. However, the way in which they behave does have consequences for endocrine status, including that of their gonadal hormones. Thus, on ascending the phylogenetic scale, we find not only that gonadal hormones permit rather than determine sexual behaviour, but that behaviour has itself become an environmental variable contributing to the regulation of reproduction.

Maternal behaviour

Parental behaviour is amenable to the same sociobiological rules that apply to sexual behaviour. Parental care is biologically costly; there must be a benefit. Ultimately the benefit is the propagation of an individual's genotype. But the question arises as to why some species minimize their investment in any one offspring for the sake of producing more, each of which is less viable, while other species produce less frequently and invest heavily in each offspring. Clearly, different strategies have been adopted in different environments and by different species. The sociobiological term for these strategies is the '*K*' or the '*r*' variety. '*K*' strategists play the game of survival by investing more time and energy in the production of few young while '*r*' strategists have a high reproductive capacity and survive where resources are only sporadically available but in excessive amounts. (This distinction is discussed more fully in Chapters 1 and 2.)

Although there is considerable variation between mammals, they are generally *K* strategists with respect to other vertebrates by virtue of the fact that the female suckles her young from specially developed mammary glands. The production of a milk that is sufficiently nutritious to sustain the young during the early stages of life gives the female a special role in parental care. The kind of maternal behaviour shown by mammals is nevertheless quite varied, depending upon the habitat and the condition of the young at birth. Altricial mammals are helpless at birth, both deaf and blind and severely deficient in motor control and temperature regulation. These offspring require considerable care by the mother, not only in nursing and suckling, but also by providing a warm nest, and retrieving them into this nest when they go astray. Mothers that produce altricial offspring (e.g. rats, mice, hamsters, rabbits, cats and dogs) also lick their offspring to stimulate urination and defaecation and are particularly defensive against predators.

Grazing mammals such as wildebeest and zebra live in large social groups, and are constantly on the move in search of food. It is important that their offspring should be relatively well advanced at birth in order to keep up with the herd. Here the young are described as precocial, and are able to both stand and walk soon after birth as well as maintain body temperature. It is also important that they can see and hear, especially if they are to recognize their mother, since most ungulate mothers care only for their own offspring. Maternal care rapidly becomes exclusive within hours of parturition, and any alien young that try to suckle may be violently rejected. The young begin to graze for themselves at quite an early age, and with the growth of teeth suckling is quickly terminated.

In primates the young are generally quite helpless at birth, but they can cling to the mother's fur and are capable of seeing and hearing. They may remain in close bodily contact with their mother, suckling for several months. Although mother and infant become separated from one another for longer periods of time with increasing age, the infants rush to motherly protection at the slightest threat of danger. In a number of primate species, related 'aunts' may participate in grooming and care of the young.

In contrast, those species that give 'premature' birth to their young (marsupials) show little change in their behaviour at parturition. The young animal crawls into the mother's pouch and attaches to a teat. Apart from cleaning her pouch more often, the mother shows little in the way of maternal behaviour and fails to distinguish her own young from any others at this time. The young of pouchless marsupials such as the murine opossums (*Monodelphis* and *Marmosa* spp.) have a well-developed grasping reflex at birth which is essential to their survival since the mother fails to nurse or hold her infants.

The development and course of maternal behaviour

Although true maternal behaviour starts with the birth of the infant, preparation for it occurs during the later stages of pregnancy. Experiments have shown that a female's maternal response to young animals increases gradually as pregnancy advances, although it attains really high values only shortly after birth. Females that produce altricial young start their nest building during pregnancy. At birth the young are covered in amniotic fluid which the mother licks dry, and when delivery of the placenta occurs this is eaten by the majority of mammalian species. We do not know exactly what prevents the mother – particularly the carnivore mother – from eating her young, but probably the way they move and the sounds they emit inhibit cannibalism. Nevertheless, mothers bearing their first litters, and mothers that are severely stressed by crowding, often do eat their offspring (and not only in carnivores). Detailed studies of parturition in the cat suggest that initially the mother regards the emerging fetus as an extension of her own body. During and shortly after parturition, the mother alternates between licking herself and her offspring. Gradually her attention focuses on the offspring, which provides her with many of the stimuli that were previously so attractive to her during birth.

The licking of post-parturient offspring is particularly important for the bonding of mother and infant in precocial ungulates. Sheep and goats deprived of contact with their newly born offspring show a rapid fading of maternal behaviour within only hours of separation. During this initial period of contact a discriminating bond is formed, which has much in common with the imprinting process described by Konrad Lorenz in precocial birds. It is formed at a 'critical period' and is generally exclusive, enabling the mother to identify and distinguish her own offspring from those of other mothers in the herd. In goats, cattle and sheep olfaction

Fig. 5.19. Giant panda with 6-week-old young. (By courtesy of Madrid Zoo, Spain.)

seems to be the most important cue for recognizing the offspring, thereby allowing the mother to suckle only her own young.

Maternal behaviour is closely synchronized with the behaviour of the developing offspring, gradually declining as the young develop physically and are able to survive independently. The giant panda in Fig. 5.19 looks quite relaxed with her 6-week-old youngster. In the cat a pattern of synchronized changes occurs between the mother and her kittens which is related primarily to feeding. The earliest phase is characterized by the mother-initiated feeding, and lasts for some 3 weeks. During the next 3–4 weeks the kittens take the initiative in feeding by approaching the mother, but she assists them. In the sixth to seventh week the mother begins to avoid the kittens when they approach to suckle and they are gradually obliged to find other food sources. This synchronized patterning of interactions requires continual contact between the mother and kittens. If isolated for as little as 2 weeks kittens are unable to adjust to the changes that have occurred in the mother's nursing behaviour during their absence.

Among monkeys the synchronized changes in behaviour between mother and infant occur over a considerably longer development period. Even after suckling has ended, the mother still keeps in close contact with her offspring. Enforced separation for only a brief period has profound initial effects on both mother and infant. Although adjustments are made when they are reunited, such infants show depressed play activity, and even 2 years later they exhibit fear responses to strange objects in unfamiliar environments more intensely than normal infants.

Hormones and maternal behaviour

The role that hormones play in influencing the various aspects of maternal behaviour has not yet been fully elucidated. Many contradictions exist and few generalizations can be made, partly because of species differences and partly because of incomplete knowledge. In the rabbit, mouse and hamster it has been possible to induce some components of maternal behaviour with exogenous hormones, but not in the rat. However, the presence of a substance in the blood of rats shortly after parturition has been reported to induce maternal behaviour when transferred into the circulation of virgin female rats. Plasma taken from donor rats within 48 h after parturition significantly reduced the latent time for retrieval of young by virgin rats, while plasma from non-pregnant rats was without effect. The onset of retrieving in the treated rats was also accompanied by crouching over the young, licking and nest building. Some of these signs appeared soon after injection (48 h) but the unresolved question was which hormone or mixture of hormones was concerned.

The basic work of Oscar Riddle way back in the 1940s suggested that prolactin was the primary hormone responsible for maternal behaviour in rodents. Although experimenters over the last 15 years have been unable to repeat these findings, some recent data indicate that prolactin plays an essential role in certain aspects of maternal behaviour in the rabbit. The data in the rat and mouse are not so conclusive with respect to prolactin, but oestrogen and progesterone do seem to play a role.

However, let me stress that maternal behaviour can occur in the absence of reproductive hormones. For example, if a non-pregnant female rat is given a litter of babies, she may ignore them, or attack and eat them. If she is presented with successive litters, she gradually begins to mother them, and if this treatment continues for several days, she will eventually develop a high degree of maternal behaviour. In fact such behaviour can even be induced in ovariectomized and hypophysectomized female rats. This process is called 'sensitization' and is not apparently mediated by any hormonal factor. Thus, cross-transfusion from sensitized rats does not induce maternal behaviour in non-sensitized rats, unlike the situation in post-parturient rats. In mice, the sensitization period necessary to elicit maternal behaviour is only 5 min. Non-pregnant hamster females eat newborn pups, but will show maternal behaviour to pups of 6 days of age and older.

Although maternal behaviour can occur without hormones, it would be quite wrong to assume that hormones are without influence. Non-parturient female rats require several days of exposure to pups before they display maternal behaviour, whereas similar rats primed with oestrogen have shorter latencies for the onset of maternal behaviour. Nevertheless, the fact remains that any amount of hormonal priming fails to induce maternal behaviour as rapidly and consistently as occurs following parturition. Indeed, if post-parturient rats were to experience the same delays in the onset of their maternal behaviour as occurs in hormonally primed rats, their pups would not survive. Clearly, parturition may play an integral part in the onset of maternal responsiveness. In this context, two recent studies are of considerable interest. Cort Pedersen and Arthur Prange in the United States have shown that intracerebral infusion of oxytocin induces immediate maternal behaviour in oestrogen-primed virgin female rats, while workers in France have shown that maternal behaviour can be induced immediately in hormonally primed sheep by vaginal and cervical stimulation. An intriguing possibility is that neural connections from the birth canal signal the intracerebral as well as hypophyseal release of oxytocin, which has simultaneous peripheral effects on labour and central effects on inducing maternal behaviour.

Establishment and maintenance of the mother–infant bond

While hormonal changes may be important for the initiation of maternal behaviour, they are not required for its maintenance. Ovariectomy and hypophysectomy do not inhibit the expression of maternal behaviour in post-parturient rats. Nevertheless, the period immediately after birth is critical for the establishment of the mother–infant bond, especially in precocial species where responsiveness to the young is exclusive. In altricial species, the postnatal period can still be important, and denying rat mothers any contact with their litters from the moment of birth retards their subsequent expression of maternal behaviour. This initial phase of the parent–offspring relationship is characterized primarily by a maternal response to cues emitted by the pups. If growth and development of the pups is retarded by malnutrition, maternal responsiveness is maintained for longer. Many of the cues from pups to which the mother responds are difficult to classify, for their meaning depends upon the context in which they are received. Nursing behaviour, for example, depends upon tactile stimulation. In rabbits, the doe will not assume the nursing posture until all her offspring are on the teats, and in cats the duration of nursing depends upon the number of kittens suckling. Suckling by the young in all species provides an important stimulation for lactation to proceed, by releasing both prolactin and oxytocin from the pituitary. Prolactin acts on the mammary gland to promote milk production, while oxytocin is important for the milk 'let-down' reflex (see also Book 3, Chapters 2 and 8, Second Edition). In rats, although suckling is the essential stimulus for lactation to commence, after 10 days the mother need only smell her pups for maintenance of prolactin secretion and lactation.

The amount of time a mother rat spends with her litter in the post-partum period depends in part upon her sensitivity to changes in body temperature. The duration of contact between the post-partum rat and her litter declines after the first week, because of the growing ability of pups to thermoregulate. This increases their ventral and core temperature, making it necessary for the mother to leave them in order to cool down.

Olfactory cues play a major role in the onset and maintenance of parental responsiveness in many mammals. Twenty minutes of licking a lamb is sufficient for sheep to bond with their lamb and subsequently reject aliens. Sight and sound of the lamb are not sufficient for ewes to remain maternal after parturition, whereas when they receive olfactory cues in addition, they always remain maternal. In some species the olfactory stimuli that maintain maternal responsiveness may be mediated by diet. In mice, communal acceptance and nursing of pups is common. However, mothers are likely to cannibalize pups from a different colony, unless their mothers have been fed the same diet when they will readily accept each other's offspring.

Among many mammalian species, auditory thresholds are lowest in the frequency ranges characteristic of offspring distress calls. Defensive behaviour has been observed in response to alarm calls in several species. Vocalizations of young may also serve for individual recognition, and reindeer, elephants, seals, sheep and monkeys seem able to recognize tape recordings of their own offspring. In mice, rats, hamsters and other species with altricial young, the infants emit what are to us inaudible (ultrasonic) vocalizations. These occur particularly when the infants are distressed, and result in intensive retrieval and nursing by the mother. Audible distress vocalizations by offspring in a number of species result in mothers coming to the aid or defence of their offspring, and in monkeys and man tactile body comfort is provided by cradling the infants.

Human mother–infant interactions

Unlike the rather restricted and specific signals that release maternal behaviour in animals, the human mother may pattern her maternal behaviour to almost any change in her infant's behaviour. John Bowlby drew particular attention to five infant behaviour systems that are effective in bringing about proximity of mother and child. These are suckling, clinging, following, crying and smiling, although any signs of distress from burping to pulling faces can have the same predictable effect. During the past decade our view of the infant has changed from seeing him or her as a passive, reacting organism to one that is capable of exerting considerable influence on maternal attitudes. Indeed, the infant's personality, moods and irritability probably have a major influence on human maternal behaviour, which are as important as attitudes the mother derives from experience of her own mother, the social world in which she grew up and her contact with other people's children. A baby that cries and shows an avoidance response when picked up may easily induce feelings of frustration and anxiety in the parents. Rejected in this way, parental attitudes may change, setting the scene for an enduring pattern of unsatisfactory relationships.

The human infant's recognition of its mother may initially depend on olfactory cues. Experiments have examined this possibility by testing whether infants can differentiate the smell of their own mother from that of strangers. Breast pads that the mother had been wearing and which had absorbed her odours were used as the stimulus for head turning by babies. By 6 days of age babies were turning towards their own mother's breast pad in preference to that of a stranger.

Babies are also able to direct their attentive responses to specific visual stimuli associated with the human face. It is the contention of some workers that the baby should be handed to its mother immediately after birth, since an early period of intense eye-to-eye contact may be very important in the formation of the mother–infant bond. There is certainly evidence that separation of mother and baby after premature deliveries can result in deficiencies in attachment behaviour if they are not permitted some early contact in the first few days. However, although disorders of mothering, including child abuse, increase disproportionately in situations associated with neonatal separation, this is not necessarily the outcome. More often than not a perfectly normal relationship develops between the mother and her baby, no matter how intrusive the external interventions might be. However, the practice of separating mothers from their babies for the first 2–3 days of life may have been partly responsible for the difficulties such mothers often encountered in establishing breast feeding.

Maternal and sexual behaviour are as essential to the process of successful reproduction as is fertilization itself. This chapter has outlined the different kinds of strategies that different species adopt to maximize their reproductive success. The hormonal mechanisms that both determine and limit these strategies have also been considered in some detail, as well as a brief mention of how these hormonal mechanisms themselves come under the influence of the environment in differing species. Even monkeys, which are relatively unaffected by the changing physical environment, show profound changes in hormonal responses to the changing social environment.

In evolutionary terms, perhaps the most significant development in sexual behaviour has been the emancipation of the female brain from gonadal influences in the catarrhine monkeys and apes. This has been achieved by switching the responsiveness of those parts of the brain that control receptive and proceptive behaviour from oestrogens and progestogens to androgens. The former two are produced cyclically, while androgens in the female are secreted from the adrenal in a fairly continuous manner. I emphasize the female brain, because to some extent this event had occurred earlier in the evolutionary history of male polygynous mammals, perhaps in keeping with their lack of parental investment.

Of course, the potential for permanent behavioural receptivity is not always expressed, even in captive monkeys, because females are not always attractive and have evolved other constraints on permanent genital compatibility by the development of the sexual skin swelling. Nevertheless, the fact that sexuality is now part of the 'id', to borrow a Freudian term, enables it, like aggressive behaviour, to become part of the 'deep structure' of social organization.

While moving up into the trees provided our early primate ancestors with an abundant supply of food and protection from predators, it also applied certain constraints on maternal behaviour. Pregnancy would restrict mobility, as would the birth of a large family. Natural selection's answer to these problems was to reduce pregnancy to a single birth and postpone much of growth and development to the postnatal period. Life among the branches emancipated the forelimb to become a grasping prehensile limb, a development that not only aided locomotion but permitted both the mother and infant to grasp each other. With the birth of a single infant, maternal investment increased, an investment further enhanced by delaying the onset of the next pregnancy. Increasing the interbirth level was a masterly stroke of evolutionary genius, since the biological process for maintaining maternal care (suckling) simultaneously inhibited the controlling mechanism for fertilization. Nature had evolved her own contraceptive, the repercussions of which were to be felt throughout the social evolution of man himself.

Suggested further reading

Ecology, sexual selection and the evolution of mating systems. S. T. Emlen and L. W. Dring. *Science*, **197**, 215–23 (1977).

Hormones and behaviour. J. Herbert. *Proceedings of the Royal Society of London*, **199**, 425–43 (1977).

Dominance and role – two concepts with dual meanings. R. A. Hinde. *Journal of Social and Biological Structures*, **1**, 27–38 (1978).

Vaginal stimulation: an important determinant of maternal bonding in sheep. E. B. Keverne, F. Levy, P. Poindron and D. Lindsay. *Science*, **219**, 81–3 (1983).

Hormonal interaction with stimulus and situational factors in the initiation of maternal behaviour in non-pregnant rats. A. Meyer and J. S. Rosenblatt. *Journal of Comparative and Physiological Psychology*, **94**, 1040–59 (1980).

Endocrine and sensory regulation of maternal behaviour in the ewe. P. Poindron and P. Le Neindre. *Advances in the Study of Behavior*, **11**, 76–119 (1980).

Induction of maternal behaviour in the virgin rat by lactating-rat brain extracts. J. Prilusky. *Physiology and Behaviour*, **26**, 149–52 (1981).

Sexual attractivity, proceptivity and receptivity in female mammals. F. A. Beach. *Hormones and Behavior*, **7**, 105–38 (1976).

Mammals, resources and reproductive strategies. T. H. Clutton-Brock and P. H. Harvey. *Nature*, **273**, 191–5 (1978).

Intraspecific variations in mating strategy. R. I. M. Dunbar. In *Perspectives in Ethology*, vol. 5, pp. 385–431. Ed. P. P. G. Bateson and P. H. Klopfer. Plenum Press; New York and London (1982).

Social influences on behaviour and neuroendocrine responsiveness of talapoin monkeys. E. B. Keverne, R. E. Meller and J. A. Eberhart. *Scandinavian Journal of Psychology*, Supplement **1**, 37–49 (1982).

The biological basis for the contraceptive effects of breast feeding. R. V. Short. In *Advances in International Maternal and Child Health*, vol. 3, pp. 27–39. Ed. D. B. Jelliffe and E. P. Jelliffe. Oxford University Press (1983).

Parental investment and sexual selection. R. C. Trivers. In *Sexual Selection and the Descent of Man*, pp. 136–79. Ed. B. Campbell. Aldine Press; Chicago (1972).

Neuroendocrinology of Reproduction. Ed. N. T. Adler. Plenum Press; New York and London (1981).

Human Sexuality and its Disorders. J. Bancroft. Churchill Livingstone; Edinburgh (1983).

Mate Choice. Ed. P. Bateson. Cambridge University Press (1983).

Reproductive Biology of the Great Apes. Ed. C. E. Graham. Academic Press; New York (1981).

Parental care in Mammals. Ed. D. Gubernick and P. Klopfer. Plenum Press; New York and London (1981).

Sexual Differentiation of the Brain. R. W. Goy and B. S. McEwen. MIT Press; London (1980).

Sex, Hormones and Behaviour. Ciba Foundation Symposium 62. Excerpta Medica; Amsterdam, Oxford, New York. (1979).

Perspectives in Primate Biology. Ed. E. H. Ashton and R. L. Holmes. Symposia of the Zoological Society of London. Academic Press; London (1981).

6

Immunological factors in reproductive fitness

NANCY J. ALEXANDER AND DEBORAH J. ANDERSON

Immune and inflammatory responses defend the body against foreign invaders such as bacteria, viruses and parasites. The immune component of this defence system is comprised of lymphocytes, which are produced by stem cells in the bone marrow and reside primarily in the blood circulation and lymphoid tissues (Fig. 6.1). The inflammatory response is mounted by other types of bone-marrow-derived cells, namely the polymorphonuclear leucocytes and monocytes, which are found in the circulation and tissues and have primarily a phagocytic role. The immune response has two characteristics that distinguish it from the inflammatory response: specificity and memory. When the immune system encounters a foreign substance (antigen) for the first time, it responds by developing a

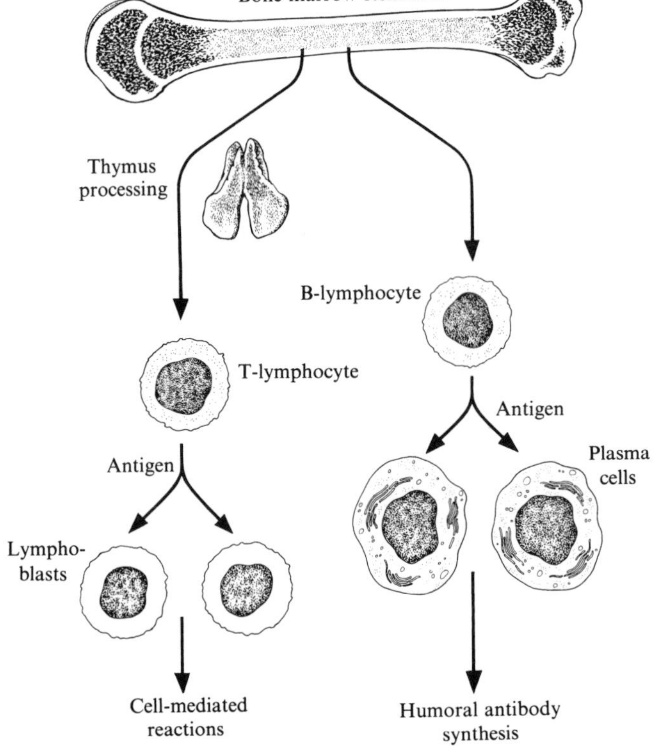

Fig. 6.1. Lymphoid tissues originate in the bone marrow. Whereas B-lymphocytes evolve directly from their stem cells, T-lymphocytes differentiate under influence from the thymus. Upon exposure to antigen, T-lymphocytes develop into specific functional types that mediate cellular immune reactions, and B-cells differentiate into plasma cells that secrete antibodies (humoral antibody responses).

reserve of specialized 'memory' cells that can recognize the same antigen in subsequent encounters. This specific memory response provides the mechanism for immunization against infectious diseases. Although inflammatory cells do not have a memory response and are not as efficient in fighting foreign invaders, they provide considerable assistance to lymphocytes during immune reactions. Macrophages (large phagocytes belonging to the monocytic series) play a key role by engulfing and processing foreign organisms or particles, and then presenting them in a highly provocative form to the lymphocytes (Fig. 6.2).

There are two major types of immune reaction: cell-mediated responses, which involve the direct participation of cells, and humoral responses, which are mediated by soluble antibodies in the circulation and secretions (Fig. 6.1). These responses are carried out by two lymphocyte classes that are distinguishable by distinct surface markers and specialized functions. The 'T'- or thymus-derived lymphocytes acquire special characteristics after undergoing maturation in the thymus. These cells are responsible for cell-mediated immune responses and play critical immunoregulatory roles as 'helper' or 'suppressor' cells. At the site of an immune reaction they secrete soluble factors (lymphokines) that potentiate the response (Table

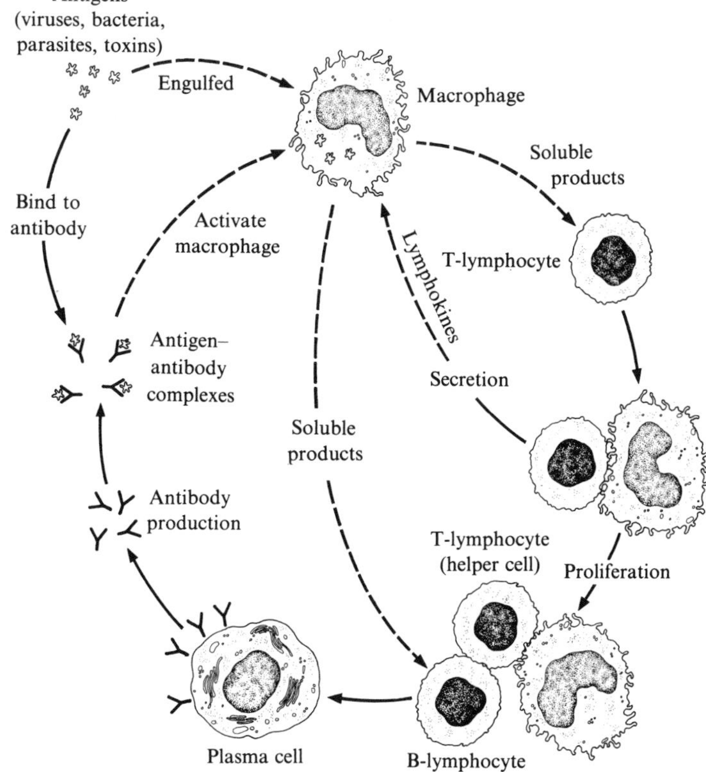

Fig. 6.2. Macrophages play an important role in immunoregulation. These cells engulf and process antigens (viruses, bacteria, parasites, toxins) and present the products to T-lymphocytes. Close apposition of the macrophages to the lymphocytes is necessary for cell–cell recognition. Some T-cells are called helper cells because they interact with B-lymphocytes and allow amplified antibody production. The B-cells subsequently differentiate into plasma cells which produce antibodies. The antibodies bind to the antigen, forming immune complexes. These immune complexes can serve to activate the macrophages. A feedback relationship thus exists with the macrophages: (1) soluble products called lymphokines produced by lymphocytes, and (2) the combined antigen and antibody – these two entities modulate the inflammatory response.

6.1). The 'B'-lymphocytes (derived from gut-associated lymphoid tissues, and the Bursa of Fabricius of the chicken – hence the B) do not undergo maturation in the thymus, but when appropriate B-cells encounter foreign antigens they differentiate into plasma cells, and it is these that then produce large amounts of antibodies. Five classes of antibodies have been described (see Table 6.2 and Fig. 6.3), each with different molecular structures and properties (Fig. 6.4). Antibodies are characterized by their ability to bind specifically and with high affinity to the foreign antigens that elicited their production. Antibodies of several classes undergo a

Table 6.1. *Some lymphokines and their properties*

Factor	Biological activity
Eosinophil chemotactin	Attracts eosinophils following interaction with immune complexes
Interferon	Has antimicrobial activity
Lymphotoxin	Kills various nucleated target cells
Lymphocyte chemotactin	Attracts other lymphocytes
Macrophage activation factor	Enhances macrophage motility and phagocytosis
Macrophage chemotactin	Attracts macrophages
Migration inhibition factor	Prevents macrophage migration *in vitro*
Mitogenic factor	Induces lymphocytes to produce more lymphocytes
Neutrophil chemotactin	Attracts neutrophils
Skin reactive factor	Produces inflammatory reaction in skin
Transfer factor	Passive transfer of delayed hypersensitivity in man

(From S. Cohen and R. T. McCluskey. Delayed hypersensitivity. In *Principles of Immunology*, p. 196. Ed. N. R. Rose, F. Milgrom and C. J. van Oss. Macmillan; New York (1973).)

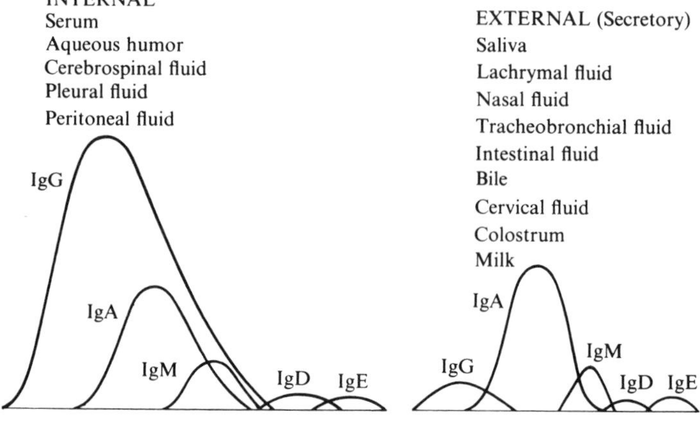

Fig. 6.3. The five immunoglobulin classes vary in concentration in the serum and external secretions. The most abundant class in serum is IgG, whereas in secretions IgA is most prevalent. The abscissa is a combination of molecular weight and electrophoretic mobility. (From T. B. Tomasi. The secretory immune system. In *Basic and Clinical Immunology*, 4th ed., ed. D. P. Stites, J. D. Stobo, H. H. Fudenberg and J. V. Wells. Lange Medical Publications; Los Altos, California (1982).)

Table 6.2. *Physical and biological properties of human immunoglobulin classes*

Class	Mean serum concentration (mg/100 ml)	Molecular weight	Half-life in circulation (days)	Biological function	Heavy chain designation	No. of subclasses	Ability to cross placenta
IgG	1210.00	150000	23	Most abundant Ig of internal body fluids, particularly extra-vascular where it combats microorganisms and their toxins	γ	4	+
IgA	280.00	170000	6	Major Ig in sero–mucous secretions where it defends external body surfaces	α	2	–
IgM	120.00	890000	5	Very effective agglutinator; produced early in immune response – effective first-line defence against bacteraemia	μ	2	–
IgD	3.00	180000	3	Present on lymphocyte surface	δ	–	–
IgE	0.03	190000	2	Raised in parasitic infections; responsible for symptoms of atopic allergy	ϵ	–	–

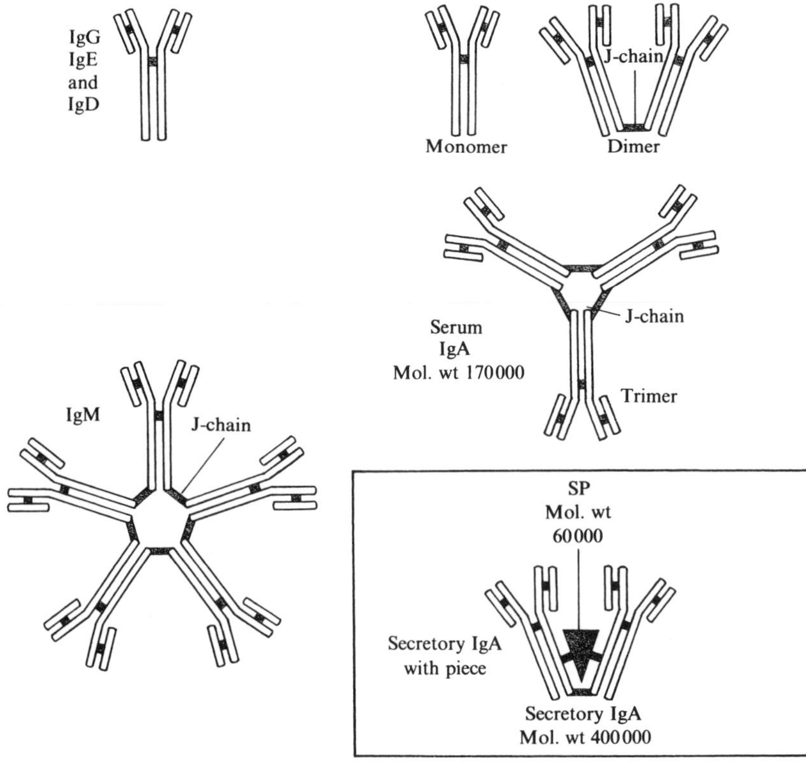

Fig. 6.4. All the immunoglobulin classes are composed of Y-shaped components. IgG, IgE and IgD are monomeric whereas serum IgA is a trimer attached by pieces called J-chains. Secretory IgA (see Fig. 6.13) is produced as a monomer by plasma cells, but then the single molecules are paired up as dimers by a secretory piece (SP). IgM, heaviest of the immunoglobulin classes, is made up of five components attached in a star-shaped arrangement by J-chain pieces. (From G. Gowland. The basic principles of immunology. In *Immunology of Human Reproduction*, pp. 1–32. Ed. J. S. Scott and W. R. Jones. Academic Press; London (1976).)

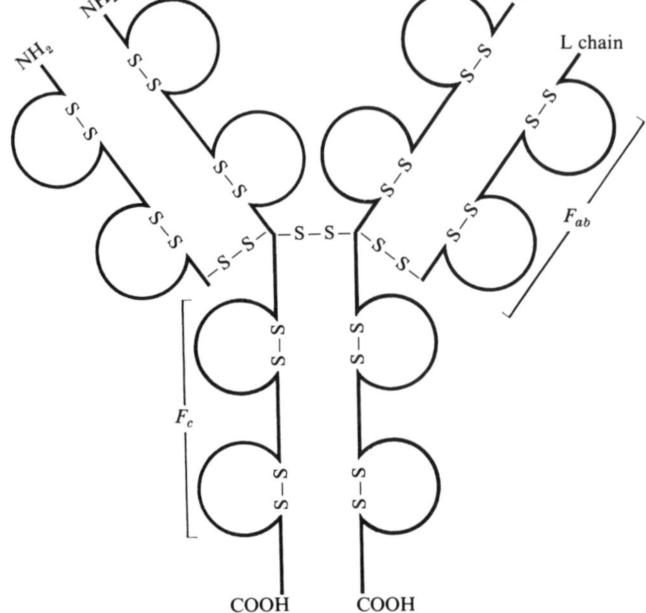

Fig. 6.5. This schematic diagram of an immunoglobulin molecule shows a central Y of two heavy (longer) polypeptide chains ('H chains') to which two light (shorter) chains ('L chains') are attached by disulphide bonds. These chains can be broken enzymatically into an F_{ab} fraction containing the antigen-binding sites and an F_c fraction which contains the complement-fixing portion of the immunoglobulin molecule. (From D. Papermaster. Immunoglobulins. In *Immunology [A Scope Monograph]*, pp. 7–17. Ed. B. A. Thomas. The Upjohn Company; Kalamazoo, Michigan (1975).)

structural change when they combine with their target antigens and thereafter can also bind serum enzymes (complement), which aid in the destruction of the antigen by digesting it or by attracting phagocytes that engulf the foreign organism or molecule. Antigen and complement binding are functions of different parts of the antibody molecule (Fig. 6.5).

Mammalian reproduction involves events and circumstances that pose some interesting immunological questions. Many elements of both the male and female reproductive systems, including the mature gametes and many specialized secretions, first appear at puberty, long after the immune system has developed and established the ability to distinguish between 'self' and 'non-self'. Therefore, these reproductive cells and secretions are 'foreign' to the immune system and can be subjected to immune attack when natural protective barriers and mechanisms break down. At each mating, the female receives millions of immunologically alien, short-lived spermatozoa, and since spermatozoa and the associated seminal fluid are highly antigenic, immune responses that could impair fertilization might be expected to develop; why this does not normally happen is something of a mystery. The developing fetus poses another intriguing immunological

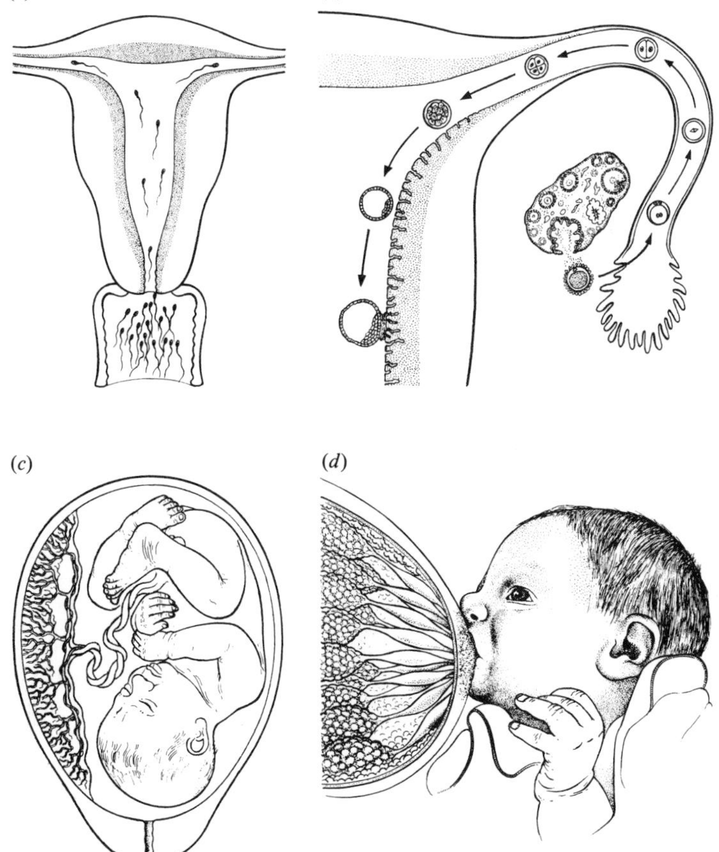

Fig. 6.6. Sexual intercourse (*a*) culminates in exposure of the female reproductive system to millions of spermatozoa suspended in a complex seminal plasma. Implantation (*b*) involves encroachment of a foreign blastocyst upon the endometrial surface. The developing fetus (*c*) achieves a parabiotic union at the level of the placenta. After parturition this relationship continues as the breast continues to provide nutrition and immunoglobulins (*d*). In some mammalian species, viable maternal leukocytes are consumed by the infant, migrate through the gut lining and gain access to the blood stream. (From A. E. Beer and R. E. Billingham. *The Immunobiology of Mammalian Reproduction*, p. 3. Prentice-Hall; Englewood Cliffs, New Jersey (1976).)

paradox: its placenta is the only known example in mammals of a naturally occurring successful foreign tissue graft. Fetal tissue expressing foreign, paternally derived, surface antigens can survive in direct contact with the maternal circulation (Figs. 6.6 and 6.7) and tissues for up to 2 years without evidence of immunological rejection by the mother. Conversely, the reproductive system can also affect the immune system. Many of the sexual secretions (such as secretions from the reproductive tract and male accessory organs) are immunosuppressive, and reproductive hormones may exert direct regulatory effects on the immune system. These hormones and immunosuppressive factors could play a critical role in protecting reproductive tissues from immunological attack. This chapter discusses situations in which immune and reproductive systems interact, and those in which the immune system impairs reproductive fitness.

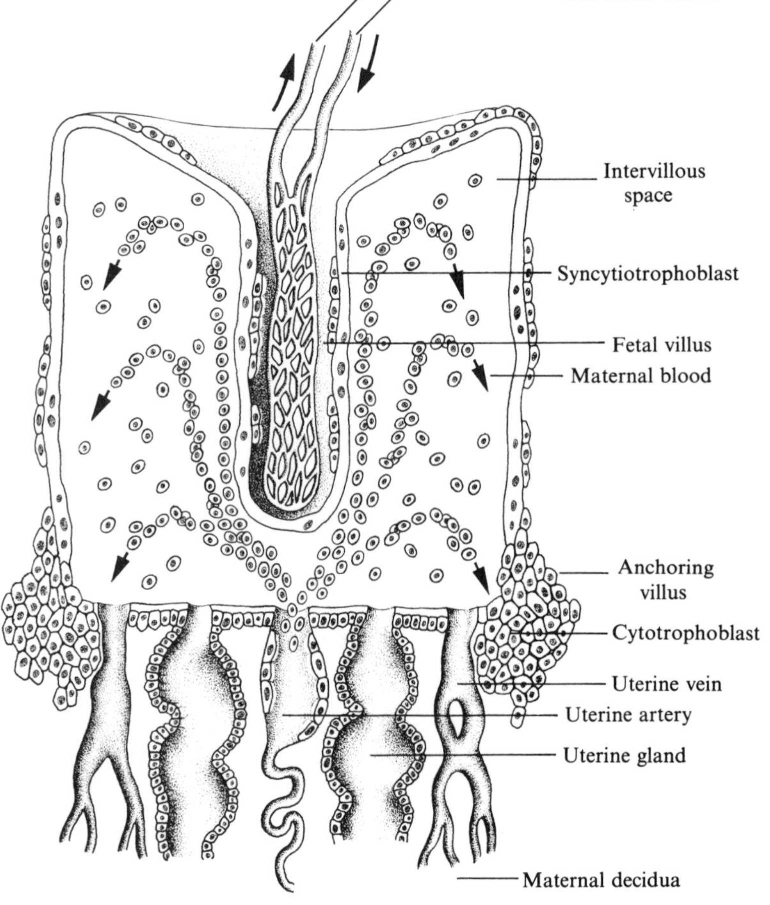

Fig. 6.7. The mature human placenta is haemochorial. The fetal tissue (chorion), consisting of three layers – syncytiotrophoblast, connective tissue (not shown), and vascular fetal endothelium (fetal blood vessels) – is in direct contact with maternal blood in the intervillous space. The cytotrophoblast of the anchoring villi is directly grafted on to the maternal decidua. Fetal blood, rich in CO_2 and metabolic waste products of the fetus, is delivered to the placenta through the umbilical arteries ('fetal blood vessels'). Blood from the villi passes via a sinusoidal network into the umbilical vein, which carries oxygenated blood back to the fetus. The oxygenated maternal blood is delivered to the intervillous space by the uterine arteries, whence it spreads out in the intervillous space as depicted and circulates among the network of vessels in the fetal villi. The venous blood returns via the uterine veins. Blood flow to the mature placenta is at the rate of about 500 ml/min. (From A. E. Beer and R. E. Billingham. *The Immunobiology of Mammalian Reproduction*, p. 17. Prentice-Hall; Englewood Cliffs, New Jersey (1976).)

The fetal allograft paradox

It is now well known that throughout differentiation and development the fetus and its placenta produce a variety of antigens that are foreign to the maternal system. During the early stages of embryonic development, unique differentiation antigens appear that have yet to be understood functionally or characterized biochemically. Shortly after implantation, paternal antigens are expressed in high concentrations on cells of the inner cell mass, which (as described in Book 2, Second Edition) is destined to become the fetus. These antigens are referred to as transplantation or histocompatibility antigens because they are of a kind first detected by tissue grafting experiments; they are capable of evoking a powerful immune response from the mother, continue to be found in high concentrations on the fetus throughout pregnancy, and occur also on cells in the placental stroma and non-villous trophoblast. Trophoblast cells have unique endocrine and transport functions that involve trophoblast-specific proteins, and in addition to expressing paternally derived gene products such as histocompatibility antigens, trophoblast cells also express tissue-specific antigens that are foreign to the maternal immune system (Fig. 6.8).

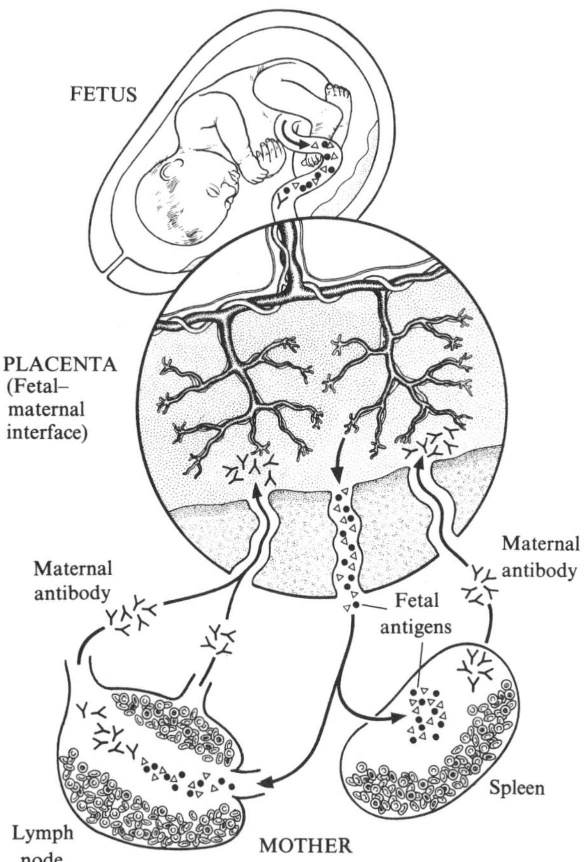

Fig. 6.8. Schematic diagram of the fetal–maternal immunological relationship. Many fetal and trophoblastic antigens are foreign to the maternal immune system and can stimulate a humoral immune response. The placenta serves as a selective filter which absorbs maternal antibodies directed against fetal antigens (harmful to the fetus), but allows passive transfer of maternal IgG antibodies (directed against antigens other than those of the fetus) which provide immunological protection to the fetus and newborn.

Peter Mẹdawar proposed three theories to explain the immunological success of viviparity: (*a*) fetal tissues may be antigenically immature, (*b*) physical barriers could exist between fetal tissue and the maternal immune system, and (*c*) maternal immunological responsiveness could be altered during pregnancy. Early experiments attempted to determine whether the

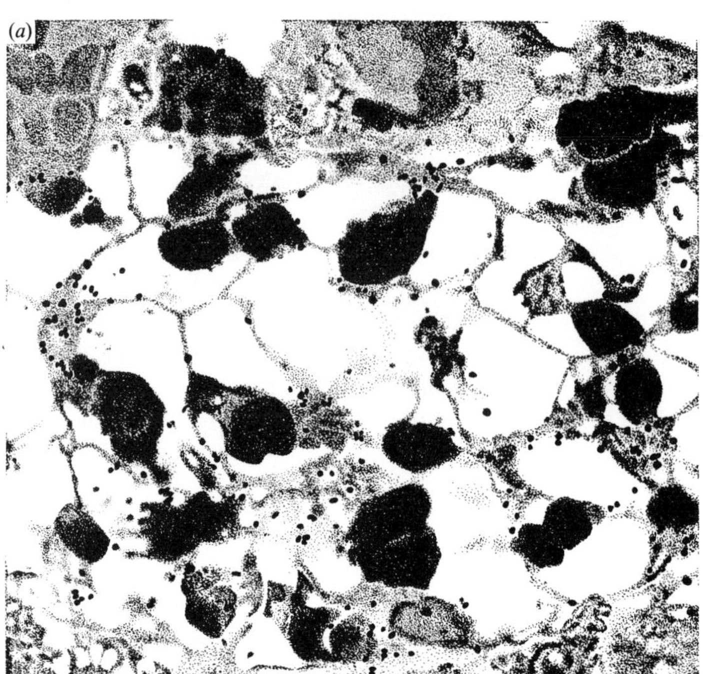

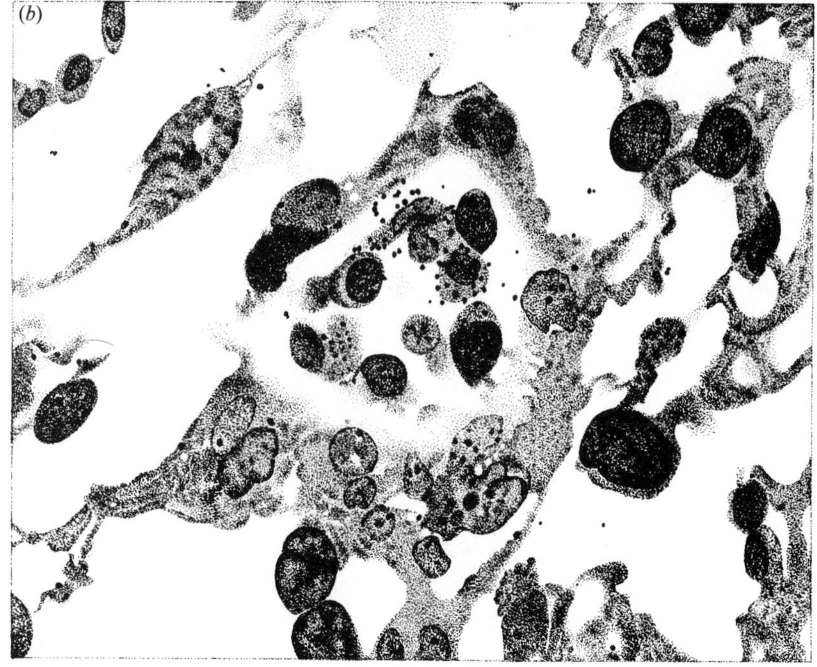

Fig. 6.9. Two photomicrographs demonstrating that the mouse placenta acts as an 'immunosorbent filter' to prevent passage of potentially harmful maternal antibodies to the fetus. Eight hours prior to the removal of the placenta shown here, the pregnant mouse received an intravenous injection of antibodies with specificity for foreign (paternally derived) fetal histocompatibility antigens (similar to the maternal antibodies induced by sensitization to the fetus during pregnancy, and which could damage the fetus if they crossed the placenta). For this experiment the antibodies had been treated with papain to remove their F_c regions (which could form non-specific placenta binding), and labelled with the radioisotope ^{125}I, so their binding site within the placenta could be detected by radioautography. Dark grains in the emulsion coating the placenta section correspond to antibody molecules. Here we can see that spongiotrophoblast (*a*) and monocytic cells (*b*) in the mouse placenta specifically bind anti-fetal antibodies and prevent their passage into the fetal circulation (From B. Singh, R. Raghupathy, D. J. Anderson and T. G. Wegmann. The placenta as an immunological barrier between mother and fetus. In *Immunology of Reproduction*, pp. 229–50. Ed. T. G. Wegmann and T. J. Gill, III. Oxford University Press; New York (1983).)

uterus was an immunologically privileged site that allowed foreign grafts to grow without eliciting an immune response. Such privileged sites are known to exist (e.g. the anterior chamber of the eye and the hamster cheek pouch), and they owe their status to a lack of lymphatic drainage (lymphatic vessels convey antigens to lymphatic tissues, where an immune response is generated). However, it was quickly shown that the uterus has adequate lymphatic drainage (Fig. 6.8) and that transplants of foreign tissue into the uterus are immunologically rejected in a classical fashion. Recently, some investigators have speculated that the decidual response in the uterine lining at the implantation site changes the lymphatic drainage and confers an immunological privilege, but this hypothesis has yet to be tested rigorously.

The developing embryo is totally surrounded and isolated from maternal blood and tissues by the trophoblast cells and the outer fetal membranes. Trophoblast cells are of embryonic origin and are in contact with maternal blood over a large surface area; it is across these cells that nutrient, excretory and gaseous exchange takes place between the fetus and mother. Trophoblast cells apparently act as an immunological barrier because they are relatively antigenically inert. They have been shown to express at least low levels of foreign (paternally derived) histocompatibility antigens when grown in tissue culture, but the antigens are coated or masked by other types of molecules, perhaps maternal proteins, when *in vivo* (Fig. 6.9). Various substances, including fibrin, sialomucin and maternal immunoglobulin, have been found to be associated with placental cells and have been implicated as antigen-masking substances. One interesting example of trophoblast cells that are immunologically rejected by the mother during the course of pregnancy is the endometrial cup tissue in the pregnant mare (see Book 3, Chapter 7, Second Edition). These fetally derived cells leave their placental site of origin and migrate deep into the endometrium, forming ulcer-like structures on its surface. Here they secrete chorionic gonadotrophin (eCG), and elicit a maternal immune response which ultimately results in their rejection by the mother. The dead tissue is sloughed off into the uterine lumen. Meanwhile, the rest of the placental tissue continues to function normally, and for unknown reasons is spared this immunological attack (Fig. 6.10).

There is considerable evidence that the maternal immune system is altered during pregnancy (Figs. 6.7 and 6.8). Various hormones, many with known immunosuppressive effects (i.e. oestrogens, progesterone, corticosteroids and gonadotrophins) are prevalent in extremely high concentrations in maternal blood during pregnancy, owing to abundant hormone synthesis particularly by the placenta. The evidence that these elevated hormone levels are associated with partial suppression of the maternal immunological system includes the following: (*a*) atrophy of maternal lymphoid tissues (particularly the thymus), (*b*) reduced reactivity of maternal lymphocytes in tissue culture, especially in assays of T-cell function, (*c*) higher maternal susceptibility to infections, and (*d*) lessening of symptoms of autoimmune diseases – all of these, during pregnancy. The immunosuppressive effects of these hormones may be quite strong near the source of their production, the placenta, where they are present in the highest concentrations. There is also evidence that trophoblast cells of the placenta secrete non-hormonal immunosuppressive factors with short-range effects. It thus seems likely that the maternal immune system is suppressed near the implantation site in the uterus. Evidence is also accumulating that certain of the gestational hormones induce lymphocyte migration, which could affect systemic immunological defence networks. For example, prolactin receptors have been demonstrated on lymphoid cells and B-lymphocytes migrate to the mammary gland during pregnancy.

Another hypothesis is that the mother may be specifically unresponsive to fetal antigens, whilst retaining close to full reactivity to all other classes of antigens. Three mechanisms have been proposed to account for this: immunological tolerance, immunological enhancement and the production of specific suppressor T-cells.

Immunological tolerance is a state of unresponsiveness to a specific antigen, usually induced by prior exposure to that antigen. The most common form is 'high-zone' tolerance, caused by administration of large doses of antigen over prolonged periods (a situation similar to that occurring during pregnancy). 'Low-zone' tolerance can be induced by

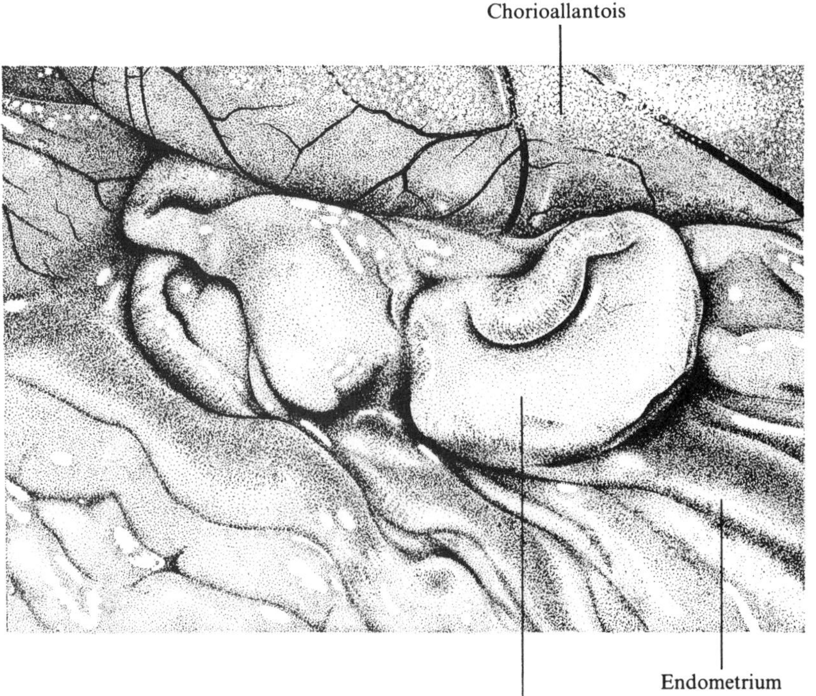

Fig. 6.10. Appearance presented by endometrial cups in a horse conceptus at 85 days of pregnancy. The allantochorion is above and the allanto-chorionic girdle below. The structure of the 'cup' is described and illustrated in Fig. 7.15 in Book 3. (By courtesy of Dr W. R. (Twink) Allen.)

continuous administration of small amounts of antigen after a strict immunization procedure. The evidence that tolerance occurs during pregnancy is scanty but tantalizing. Studies both *in vivo* and *in vitro* suggest that the maternal immune response is less vigorous towards paternal antigens than towards unrelated (third-party) histocompatibility antigens. The possibility of specific tolerance towards other fetal antigens should also be investigated.

The second mechanism of specific immunological protection that may be operative during pregnancy is called *immunological enhancement*. This effect was first described in tumours, when tumour growth was shown to be enhanced in animals previously immunized against specific tumour antigens. The enhancement mechanism is mediated by specific 'blocking' antibodies that interfere with the normal cell-mediated immune destruction of the tumour. These blocking antibodies work in three ways. In 'efferent' enhancement non-cytotoxic antibodies bind to foreign antigens on the target cell and mask their antigenicity, so that these same foreign antigens can no longer be recognized and destroyed by cytotoxic T-cells. In 'central' enhancement blocking antibodies bind to soluble antigens and interfere with their circulation to lymphoid tissues, thereby preventing sensitization to these antigens. In 'afferent' enhancement antibodies or antibody–antigen complexes bind to antigen receptors on effector lymphocytes and block their function. There is good evidence that efferent enhancement occurs during pregnancy. Considerable amounts of maternal antibody are associated with trophoblast cells in the placenta, and these antibodies are trophoblast-specific and primarily of non-complement-fixing (non-cytotoxic) classes. To date evidence that central or afferent enhancement also occurs during pregnancy is not convincing.

A third proposed mechanism to account for specific alterations in maternal immunological responsiveness is action by *suppressor T-cells* to inhibit both cell-mediated and humoral responses. Such suppressor cells have been found in the lymph nodes draining the fetal implantation site and also in umbilical cord blood.

Passive transfer of immunity from mother to offspring

In mammals the immune system is underdeveloped at birth. The cell-mediated (T-cell) system develops early in fetal life in most species, but is primitive and ineffective because the fetus lives in a sheltered environment and does not usually encounter infectious agents. The humoral immune system develops much later (after birth in many species) and also does not initially function adequately owing to lack of exposure to foreign antigens. The protection that the new-born needs for a few days or weeks while its own immune defences develop is conferred by maternal antibodies and lymphocytes which are transported across the placenta or other fetal membranes during pregnancy, or are transferred to the neonate through colostrum and milk (Fig. 6.6).

The route, timing and type of immunity transferred from mother to offspring vary considerably from species to species (Table 6.3). In the rabbit the majority of maternal antibody is transmitted across the fetal membranes during pregnancy. Antibodies are transported from the maternal circulation into the lumen of the yolk sac cavity, where they pass into the vitelline circulation and thus directly to the fetus (Fig. 6.11). This is made possible because part of the yolk sac wall breaks down during placentation (see Book 2, Chapter 2, Second Edition). This species has been used as a model for studies of the mechanism of maternal–fetal antibody transport. Experiments with enzyme-digested antibody fragments have shown that

Table 6.3. *Amount of maternal–fetal and maternal–neonatal antibody transfer in various species*

	Placental (IgG)	Colostrum (IgG)	Milk (IgA)
Pig	0	+++	+++
Horse	0	+++	?
Sheep, cow	0	+++	+++
Cat, dog	+	++	+++
Rat, mouse	+	++	+++
Rabbit, guinea-pig	+++	±	+++
Man, monkey	+++	±	+++

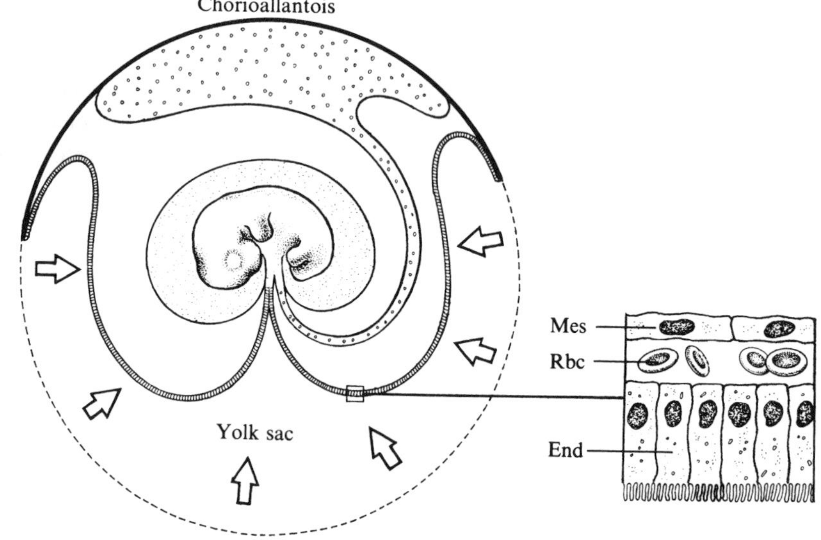

Fig. 6.11. Diagrammatic representation of the inverted yolk sac placenta of the rabbit. Immunoglobulins present in uterine secretions (represented by arrows) are transported across the yolk sac to the fetus. In most non-primate mammals, transport of immunoglobulin occurs across intra- or extra-embryonic epithelia of endodermal origin, but in humans antibodies pass through the ectodermal epithelium of the trophoblast. Mes, mesoderm; Rbc, red blood cells; End, endoderm. (From D. H. Stevens and C. A. Samuel. The anatomy of placental transfer. In *Placental Transfer*, pp. 1–14. Ed. G. V. P. Chamberlain and A. W. Wilkinson. Pitman Medical; London (1979).)

segments of antibodies containing the F_c region (Fig. 6.5) are transported very quickly to the fetus, whereas antibody fragments containing only the F_{ab} region (the antibody combining region) are not transported. More recently it has been shown that antibody molecules are actively transported within vesicles across fetal membranes after attachment to F_c receptors on the surfaces of cells with transport capabilities.

In the cow and pig, transport of IgG antibodies occurs exclusively through the colostrum within a few hours of birth. Milk from various species can contain quite different immunoglobulin profiles. In humans, the predominant immunoglobulin class in colostrum is secretory IgA, whereas in cows it is IgG. Antibodies of all classes, apparently originating from the serum, are concentrated in the colostrum (rather than mature milk) (Table 6.4) and are efficiently transferred to the newborn through the gut epithelium. However, the neonatal gut in these species is only transiently hospitable to maternal antibodies; within a few hours it undergoes a pH change and digestive enzymes appear that can destroy antibodies.

In rodents, whose offspring are extremely underdeveloped at birth, transfer of maternal antibody occurs via the placenta before birth, and via the colostrum and milk for several days after birth. In these species antibodies transported across the placenta are restricted to certain IgG subclasses. Several antibody classes appear in the colostrum, but there is a subsequent shift to greater concentrations of IgA antibodies after the first post-partum day. These IgA antibodies are synthesized locally in the mammary tissues and are joined together in dimers by a short peptide chain that stabilizes the molecules in secretions (the dimers are called S-IgA [secretory-IgA]) (Fig. 6.12). These maternal S-IgA molecules show activity against a variety of microorganisms, especially the *Escherichia coli* group, and are thought to be important in antimicrobial protection of mucosal surfaces in the newborn. Production of secretory immunoglobulins involves

Table 6.4. *Comparison of immunoglobulin content of human and cows' milk*

	Human (mg/ml)	Cow (mg/ml)
Colostrum		
Secretory IgA	5.00	10.0
IgG	0.15–0.70	60.0
IgM	4.00	10.0
Mature milk		
Secretory IgA	2.00	0.5
IgG	0.56	1.0
IgM	0.15–0.70	0.5

several steps. Specialized regions of the gut called Peyer's patches contain lymphoid cells (GALT or gut-associated lymphoid tissue). These lymphoid cells become activated by intestinal antigens or non-specific intestinal factors (e.g. bacterial lipopolysaccharides). Exposure to these antigens takes place in specialized epithelial cells called M-cells. IgA-bearing lymphoblasts then migrate to the mesenteric lymph nodes, enter the circulation, and finally home in on the gut epithelium, and the salivary and mammary glands (Fig. 6.13). In humans, passive immunity is derived almost exclusively from placental transport of maternal antibodies before birth. Only IgG crosses the human placenta, and transfer begins as early as week 12 of gestation. Evidence concerning the transmission of immunity to the human newborn is conflicting. Secretory IgA is found in high concentrations in human breast milk, but is not transported readily across the intestinal wall because of its dimer form; however, it plays an important role in local defence of the gut mucosal surfaces.

Human and rodent mammary secretions contain a variety of cells, including a large number of lymphocytes and monocytes. Their role in human immunity has yet to be established, but in mice these cells are fully functional and are transferred to the newborn via the gut. As in the case

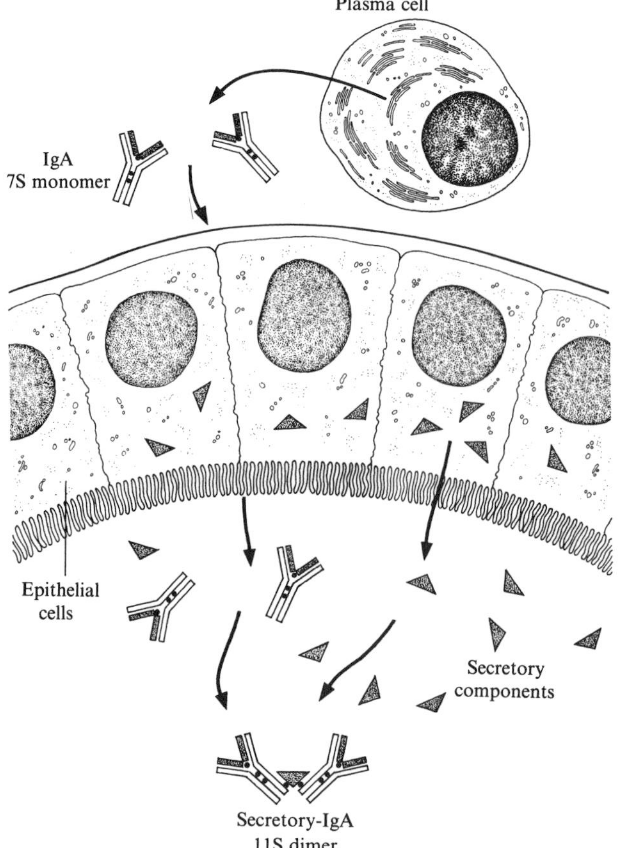

Fig. 6.12. Monomeric IgA is secreted by plasma cells, while a secretory component is produced by epithelial cells. This component binds two IgA monomers together into an 11S dimer. (From J. A. Bellanti. *Immunology*, p. 105. Saunders; Philadelphia (1971).)

of maternal immunoglobulins, the neonatal gut is permeable to maternal cells for only a short period after birth. Some researchers speculate that these cells help the neonate establish cell-mediated and immunoregulatory functions, but we still cannot describe the specific mechanisms.

Immunopathology of reproduction

Several examples exist of adverse immunological effects on reproductive processes. The best known example is Rh sensitization, in which a rhesus-antigen-negative (Rh-negative) mother produces antibodies against Rh-positive red blood cells of her own fetus (Fig. 6.14). High titres of anti-Rh antibodies are generally produced only after the first delivery, when extensive haemorrhage of fetal blood cells into the maternal system occurs. Thereafter, maternal anti-Rh antibodies are transported across the placenta in subsequent pregnancies (by normal placental transport mechanisms), and can destroy fetal red cells bearing the Rh marker and induce a serious disease called erythroblastosis fetalis or haemolytic disease of the newborn. Thanks to the pioneering work of Cyril Clarke in Liverpool, physicians now treat Rh-negative women with large doses of anti-Rh antibodies at delivery; these block Rh antigen sites on any fetal blood cells that leak into the maternal circulation, and hence prevent maternal sensitization. Prior to this practice fetal death occurred in more than 10 per cent of the pregnancies of women previously sensitized against

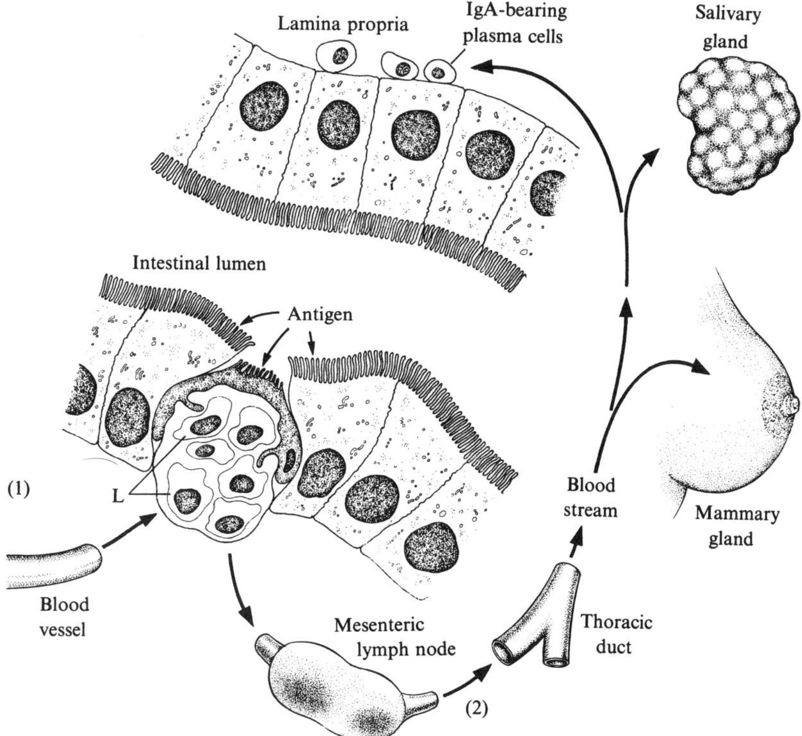

Fig. 6.13. The entero-mammary circulation. (1) Lymphocytes (L) within gut-associated lymphoid tissue, primarily the Peyer's patches of the ileum, are stimulated by antigens in the intestinal lumen. (2) These stimulated lymphocytes migrate to mesenteric lymph nodes where they further mature, and then enter the systemic circulation to home in on, amongst other tissues, the mammary gland, salivary glands and gut epithelium, where they become plasma cells and secrete IgA. (From W. A. Walker and K. J. Isselbacher. Intestinal antibodies. *New England J. Med.* **297**, 767–73 (1977).)

fetal rhesus antigen, and many liveborn infants had jaundice or permanent brain damage.

Other fetal antigens can stimulate maternal immunological responses, but usually without serious consequences. Large amounts of antibodies directed against paternal histocompatibility antigens are often found in pregnancy serum samples, but are apparently not transported to the fetus (where they would no doubt cause extensive damage). Mouse studies have shown that trophoblast cells express just enough paternal histocompatibility antigens to enable the placenta to filter out such antipaternal maternal antibodies and prevent their transplacental passage to the fetus. Animal studies have shown that it is very difficult to induce abortion even by intentional immunization with large amounts of fetal histocompatibility antigens. Nonetheless, it is possible that natural abortion in some instances is due to maternal immunological reactions to fetal products, and immunological mechanisms may be a particularly important and effective protection against developmentally abnormal fetuses, which express highly unusual new antigens or old antigens in new ways.

The zona pellucida surrounding the oocyte (described in Chapter 3 of Book 1, Second Edition) contains proteins that are unique and potentially autoantigenic to the female. However, they rarely elicit immunological responses, probably owing to the small amount of tissue present at one time, and the fact that the Graafian follicle is an extra-vascular site. Spermatozoa are a more serious potential problem to female reproductive fitness. Sexually active females are immunized regularly with millions of

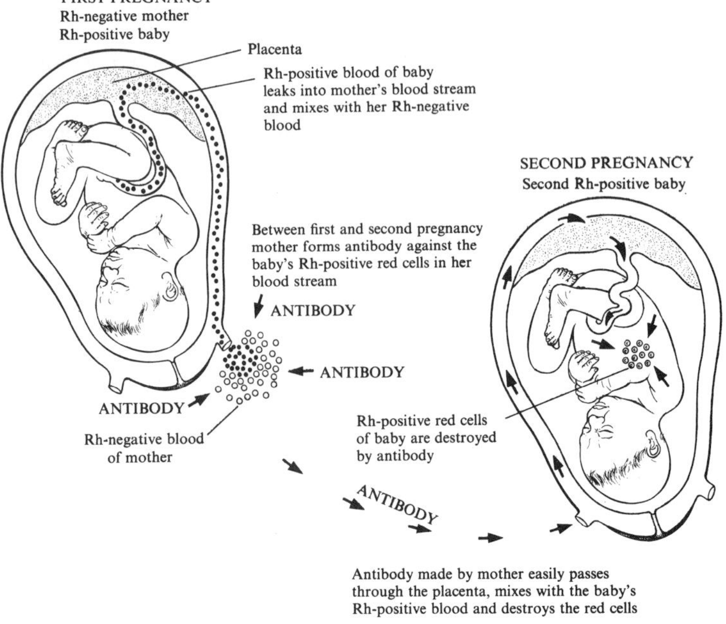

Fig. 6.14. If a Rh-negative mother is carrying a Rh-positive fetus expressing one of the Rh antigens on its erythrocytes, she will produce anti-Rh antibodies if fetal Rh-positive blood enters her circulation; this commonly happens at the time of parturition. If there is a subsequent pregnancy, transmission of these antibodies to the fetus across the placenta will result in destruction of fetal erythrocytes and development of haemolytic disease of the newborn, with fetal growth retardation, jaundice, and even death. (From A. E. Beer and R. E. Billingham. *The Immunobiology of Mammalian Reproduction*, p. 151. Prentice-Hall; Englewood Cliffs, New Jersey (1976).)

spermatozoa, and occasionally develop severe immunological sensitization to these cells or to seminal secretions. Prostitutes commonly have high antisperm antibody titres. Spermatozoa express a variety of foreign antigens, and antibodies and cell-mediated immune responses developed in response to these antigens interfere with sperm function and can cause infertility (see the detailed discussion later in this chapter). A few cases have been reported in the medical literature of women experiencing severe systemic anaphylactic shock reactions because of immunological sensitization to semen after intercourse, but these situations fortunately are quite rare.

We have recently discovered that immunity to spermatozoa occurring after vasectomy could have serious systemic effects as well as local effects in the testis. When high titres of antibody and large amounts of antigen are both present (spermatozoa continue to be produced after vasectomy, so there is much antigen constantly present), 'immune complexes' (precipitates of combined antigen and antibody) can form in the circulation. If the complexes are large they are rapidly removed from the circulation by phagocytes, but if they are small they continue to circulate and since they bind complement (lytic enzymes) they can cause extensive vascular injury. Studies in rabbits and in rhesus and cynomolgus monkeys have shown a correlation between vasectomy and atherosclerosis, but so far there is no evidence of this in men.

Effects of gonadal hormones on immune functions

Many hormones, particularly steroid hormones, exert modulating influences on certain immunological functions. Corticosteroids produced by the adrenal glands induce lymphocyte death and atrophy of lymphoid tissues in many species. Many sex hormones, such as oestrogens, progestins and androgens, also affect the lymphoid tissues and immunological responsiveness, although the effects of physiological levels of these hormones are more subtle than those produced by adrenal hormones. Effects of gonadal hormones are more apparent in males than females. It has been shown that castration of males slows down the natural involution of the thymus with age, and that testosterone injection speeds the process up. Castrated males are slightly more immunologically competent than normal intact males of the same age, and non-pregnant females are more immunologically responsive than intact males, probably because they do not maintain such high levels of circulating gonadal hormones. 'New Zealand' mice (*NZB/NZW*) have provided an interesting model for studies on the effects of gonadal hormones on immune responses. In this hybrid mouse strain severe autoimmune disease develops: the cells of the immune system forget what is self, attack other normal cells of the body and eventually cause death. Although the symptoms are caused by damage inflicted by autoantibodies, the central problem appears to be aberrant immunoregulation, probably attributable to a T-cell defect. Females have an earlier onset of autoimmune symptoms and a much shorter life expectancy than males; however, the disease develops in castrated males as early as it does in females, and testosterone-treated females show delayed onset of the disease. It appears that testosterone exerts a protective immunoregulatory effect, although we do not know how this happens.

It has been postulated that immunosuppressive reproductive hormones, present in very high levels near their sites of synthesis in the gonads, play a critical role in suppressing immune responses against gametes.

Autoimmune orchitis

Since many proteins and other molecules associated with the male reproductive tract are unique and are not produced until puberty, the immune system recognizes them as foreign and can respond to them. Spermatozoa and other products of the reproductive tract are normally sequestered from the immune system by physical barriers formed by tight junctions between Sertoli cells (Fig. 6.15) lining the seminiferous tubules and by epithelial cells lining distal portions of the tract. (The barrier formed by the Sertoli cells in the testis is called the blood–testis barrier, and is one

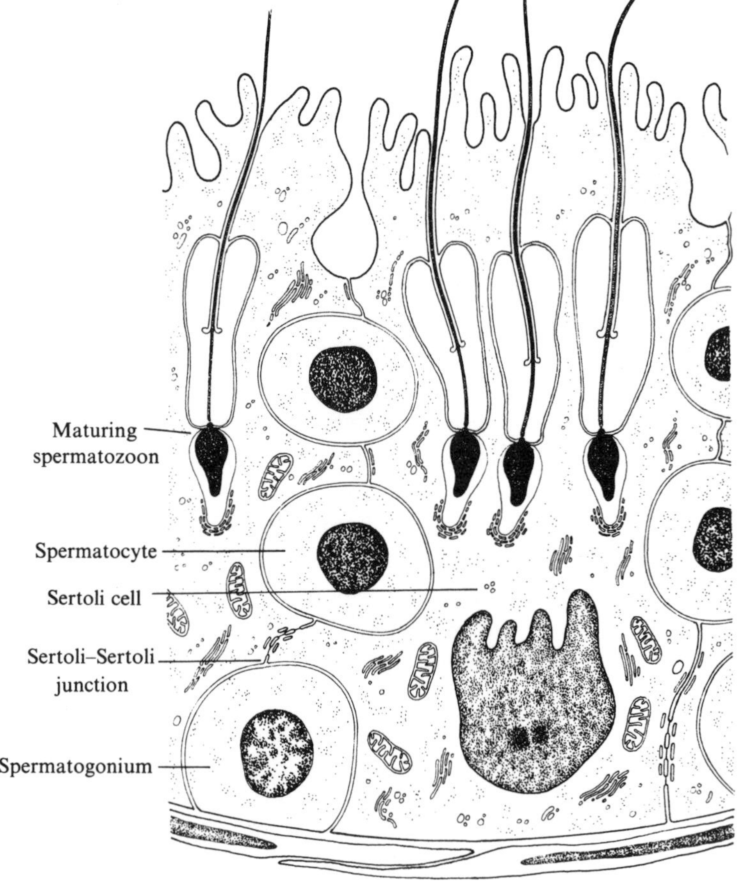

Fig. 6.15. Seminiferous tubules are composed of germinal epithelium (spermatogonia, spermatocytes and spermatids) and Sertoli cells. Sertoli cells aid in sperm maturation, actively phagocytose degenerating cellular material, and their tight junctions provide a blood–testis barrier which sequesters maturing spermatozoa from the immune system.

of the strongest tissue barriers in the body – further details are given in Chapter 4 of Book 1, Second Edition). However, even with such a barrier, an immune response to spermatozoa often occurs after testicular injury, disease or obstruction of the reproductive tract (e.g. testicular biopsy, mumps or vasectomy) (Fig. 6.16).

To study this autoimmune phenomenon, researchers have induced experimental allergic orchitis by immunization with homogenized testis or spermatozoa (Fig. 6.17). Studies in a variety of species reveal that both cell-mediated and antibody responses develop about 6 weeks after immunization with antigen suspended in a vehicle known as complete Freund's adjuvant, and exfoliation of germinal cells occurs within the testis. Spermatids, spermatocytes and even spermatogonia may be eliminated, depending upon the immunization schedule, and fertility can be seriously impaired.

Allergic orchitis does not occur frequently in normal individuals because extremely effective mechanisms deal with the prodigious quantity of by-products of spermatogenesis. One such mechanism is the rapid phagocytosis and degradation of sperm products by Sertoli cells, which have multiple functions in addition to the formation of the blood–testis barrier.

The events that occur after immunization to cause orchitis are not fully understood. If the seminiferous tubules are immunologically privileged sites (regions not accessible to the immune system), then what causes the breakdown of the blood–testis barrier allowing the initial lymphocytic infiltration into the testis? Allergic orchitis does not develop in prepubertal guinea-pigs immunized with adult testis plus Freund's adjuvant, so it appears that the disease does not occur until the germinal epithelium and the associated tight junctions are fully developed. The rete testis and efferent ducts may be the weak area in the blood–testis barrier system;

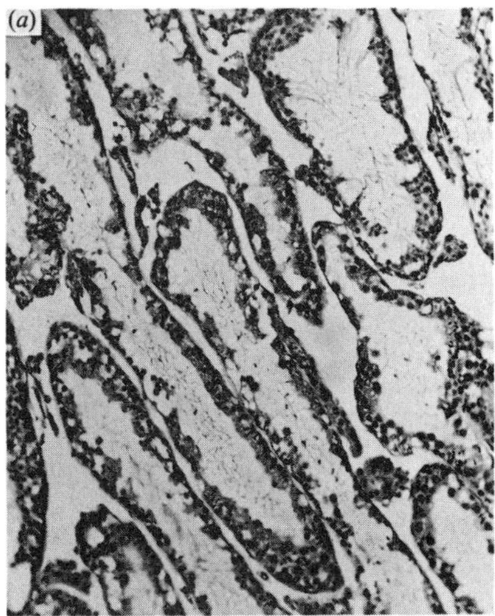

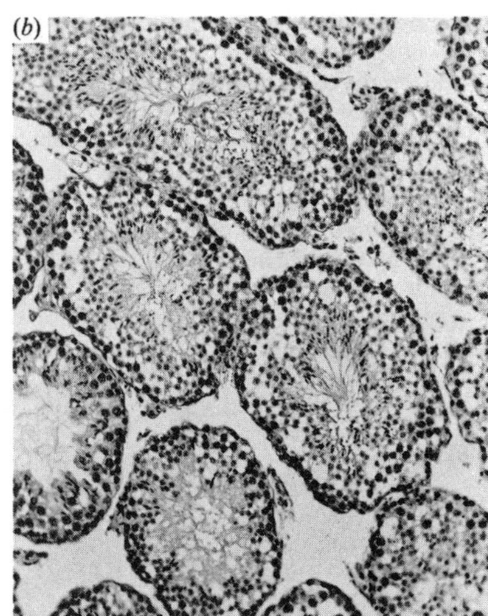

Fig. 6.16. Allergic orchitis results in loss of first the mature spermatozoa and then spermatocytes and finally spermatogonia. In some species reduced spermatogenesis also occurs after vasectomy. Compare the section of testis from a mouse vasectomized 15 months previously (*a*) with that from a sham-operated control (*b*). (From D. J. Anderson and N. J. Alexander. Antisperm antibody titres, immune complex deposition and immunocompetence in long-term vasectomized mice. *Clin. Exp. Immunol.* **43**, 99–108 (1981).)

antibodies and lymphoid cells may enter at these sites and migrate into the seminiferous tubules and epididymis (Fig. 6.15). There is evidence that sperm antigens are only partially sequestered in the efferent ducts and that soluble products leak out. This leakage could have physiological advantages since traces of antigen could induce low zone tolerance or activate certain types of T-lymphocytes called suppressor cells. The recent finding that isolated testicular germ cells can suppress lymphocytic function *in vitro*, whereas somatic cells from the same testis could not, supports this hypothesis of tolerance.

T-cells play an important role in the pathogenesis of allergic orchitis. Lymphoid cells collected from the peritoneal cavity of sperm-immunized guinea-pigs produce a lymphokine called macrophage inhibition factor in the presence of sperm antigens; purification of the cells producing this lymphokine has shown them to be T-cells. Severe allergic orchitis can be transferred in inbred guinea-pigs by injecting recipients with T-cell-enriched populations of peritoneal or lymph node cells from donors immunized with purified guinea-pig aspermatogenic antigen (Fig. 6.17). On the other hand, transfer of antibodies from the serum of immunized donors does not usually induce allergic orchitis. This indicates that cellular immune mechanisms play a more important role in the induction of disease than the humoral components of the immune system.

The testis and spermatozoa contain many autoantigens. We are only just beginning to determine which ones are associated with the development

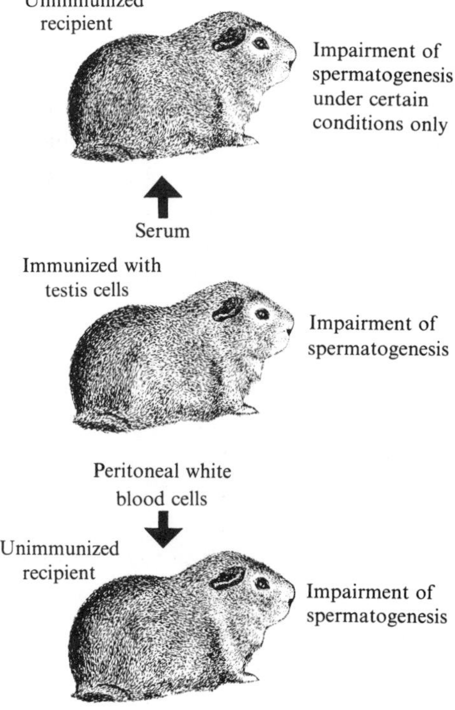

Fig. 6.17. At one time, immunologically active cells were believed to be the major cause of autoimmune aspermatogenesis. This figure shows that white blood cells transferred from an immunized guinea-pig to an untreated animal can impair spermatogenesis (lower part of figure); but serum from immunized guinea-pigs has also been shown to induce lesions under certain circumstances (upper part). (Based on Fig. 4.5 in Book 4, Chapter 4, First Edition.)

of orchitis. We know that many of the aspermatogenic antigens are proteins or glycoproteins located on the sperm acrosome. It is possible to semi-purify these antigens so that only a few microgrammes are needed for immunization.

There are several examples of spontaneously occurring orchitis. In a certain strain of beagle dogs, testicular disease closely resembling experimental allergic orchitis is commonly observed in healthy 'normal' (non-immunized) males. Testicular biopsies from these dogs show aspermatogenic seminiferous tubules with focal to diffuse infiltrates of mononuclear cells. About 40 per cent of a certain strain of mice, when heterozygous for a recessive lethal gene, exhibit severe orchitis. Focal monocytic orchitis has been found in about one-third of the testes from normal adult rhesus macaques. Such immune responses could be attributable to one of the following: (*a*) a physical defect, such as disruption or obstruction of the reproductive tract, (*b*) an abnormal immune response that is genetically determined, or (*c*) an imbalance in normal immunoregulatory mechanisms. The hypothesis that immunoregulatory mechanisms normally hold orchitic reactions in check originates from the finding that neonatally thymectomized Lewis rats develop orchitis as adults. Neonatal thymectomy results in a lack of suppressor (regulatory) T-cells later in life, which could account for the development of testicular disease in this case. Ageing also results in an increase in antisperm antibody titres, presumably for the same reason (regulatory T-cell activity decreases with age).

Vasectomy, which blocks the sperm outlet from the testis, often elicits an immune response to sperm and testicular antigens. Antibodies to spermatozoa develop as a result of vasectomy in all species, including man, but circulating antibodies are not found in all vasectomized individuals. We do not know why the response is variable but there are several possible explanations. One hypothesis is that antibody production after vasectomy is related to sperm production; individuals with low sperm production rates or small testes do not produce as much antigen and therefore do not immunize themselves as effectively as individuals with high sperm production. A second hypothesis is that sperm granulomas (walled off areas of extravasated spermatozoa) and the associated foreign body giant cells which form in tissue adjacent to the epididymis in many individuals, enhance sperm antigen presentation to the immune system. A third hypothesis is that the genetic make-up of an individual influences the type and intensity of the immune response directed towards sperm antigens. A recent study of vasectomized men revealed an association between a major histocompatibility antigen, A-28, and the presence of a type of antisperm antibody. Another possibility is that the variable immune response may be because current antibody assays detect only free antibodies, not antibodies bound to antigens. In cases where individuals have high titres of antisperm antibodies, but also produce large quantities of soluble

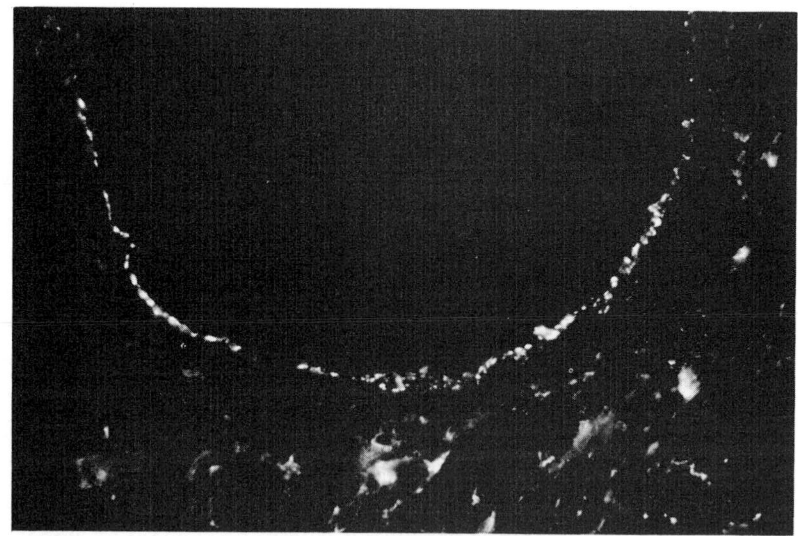

Fig. 6.18. The basement membrane surrounding a seminiferous tubule of a vasectomized rabbit shines like a string of bright beads when stained with fluorescein, indicative of the deposition of sperm antigen and antibody.

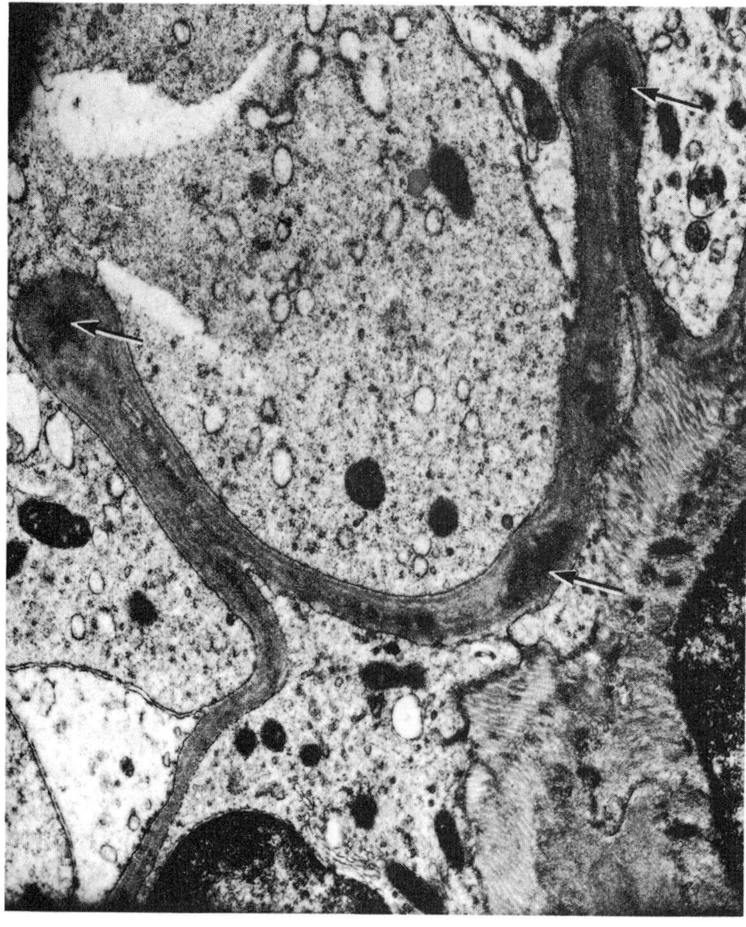

Fig. 6.19. By electron microscopy, electron densities (arrows) can be found within the basement membrane that surrounds this seminiferous tubule from a vasectomized rabbit.

antigen, antibodies may not be detected because they are bound up in immune complexes.

Since obstruction of the vas deferens can lead to autoimmune responses to testis-specific antigens, a logical question is whether vasectomy results in autoimmune orchitis (Fig. 6.16). There is no definite answer to this question at present. We know that immune complexes often develop in testes of long-term vasectomized rabbits. Granular deposits of IgG and complement are found in the basement membrane of the seminiferous tubules (Fig. 6.18). With an electron microscope, dense deposits indicative of antigen–antibody complexes can be seen in this same area (Fig. 6.19). In monkeys, more vasectomized than control animals exhibit orchitis (i.e. aspermatogenesis and lymphocytic infiltration). Immune complexes, as revealed by granular deposits of IgG and complement, are not deposited around the seminiferous tubules in monkeys, but are most commonly found in the basement membrane surrounding the ductuli efferentes and epididymides. Such findings suggest an increase in either seminiferous tubule or epididymal permeability to sperm antigens after vasectomy, depending on the species. There is conflicting evidence on testicular pathology in humans after vasectomy, some groups finding no detectable change and others reporting patchy (focal) orchitic lesions.

Immunity and infertility

It is difficult to estimate the number of couples that are infertile because of immunological factors. Antibodies may cause subfertility rather than infertility, and increase the time required for a couple to achieve a pregnancy rather than absolutely preventing it. It is difficult to quantify such reduced fertility, and even harder to evaluate it when testing the efficacy of various therapeutic measures.

We have discussed several hypotheses to explain why antisperm antibodies develop in some men. Such antibodies also develop in a small percentage of women, and the mechanisms underlying their production are also not clearly understood. As mentioned earlier, immune responses vary from one individual to another; certain women may have a genetic predisposition for a greater immune response. Since antibodies against spermatozoa do not usually develop in female animals, some people attribute antibody development in women to the fact that humans, unlike other species, mate throughout the ovulatory cycle, and claim that the uterine tissue is more reactive to foreign cells during the luteal phase of the cycle. This is based on the finding that there are many spermatozoa within uterine tissue during this period and also an influx of macrophages; these could enhance an immune response. Another hypothesis to explain the development of antisperm antibodies in women is the presence of spermatozoa during an infection of the reproductive tract, which could have an adjuvant effect. Another possibility is that cross-reacting antibodies (i.e. antibodies that are formed against one antigen but can also react with a second antigen) may develop in women and impair fertility. Studies have shown cross-reactivity between certain bacterial antigens and spermatozoa. Prostitutes are more likely to have sperm antibodies, presumably because of exposure to a large amount of sperm antigen, although the possibility that other sexually transmitted diseases have produced an adjuvant effect in these cases and have stimulated an immune response cannot be overlooked. Finally, it is possible that the seminal plasma of certain men lacks factors that normally suppress immune responses to spermatozoa in women. *In vitro* and *in vivo* immunological studies indicate that seminal plasma contains potent immunosuppressive factors that can suppress antisperm responses. Such secretions may coat spermatozoa and normally prevent the women's immune system from becoming activated.

Mechanisms underlying antibody-induced infertility are little understood. There are several tests for sperm-specific antibodies; those that cause spermatozoa to agglutinate are probably the most commonly studied. Serum from an immune individual mixed with spermatozoa will result in spermatozoa clumping in either a head-to-head, a tail-to-tail, or a mixed type of agglutination (Fig. 6.20). Several classes of antisperm antibodies, namely IgG, IgM and IgA, cause such agglutination. Because IgM has more combining sites, the agglutination is likely to be more macroscopic when this immunoglobulin is involved. Although some individuals with sperm-agglutinating antibodies may be fertile, it is safe to

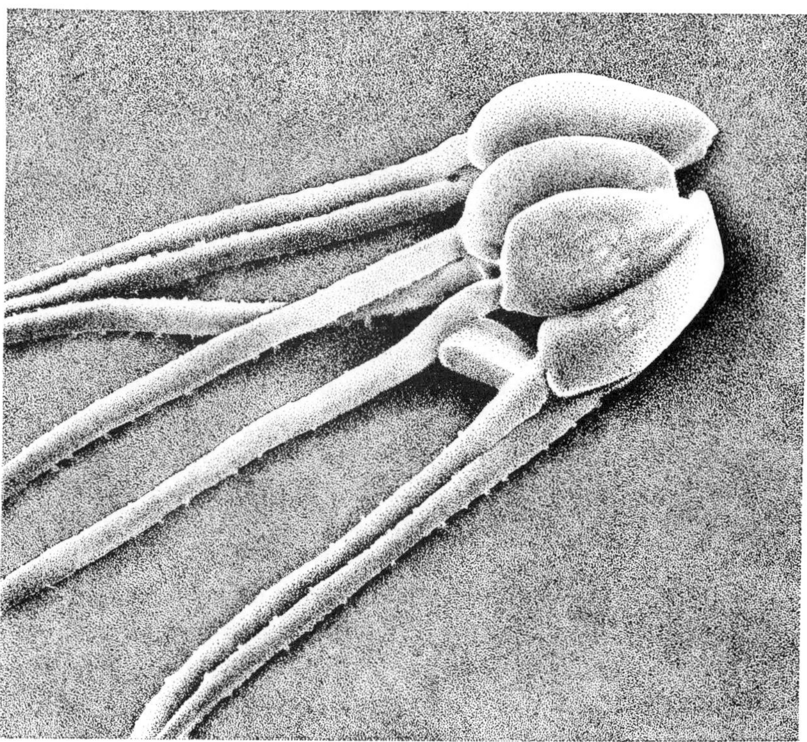

Fig. 6.20. Head-to-head agglutination of rhesus monkey spermatozoa that have been exposed to an immune serum. (Stereoscan micrograph.)

say that sperm agglutination is more common in infertile than in fertile individuals. Spermatozoa in ejaculates from men with circulating sperm antibodies often are agglutinated. Their seminal plasma also contains antibodies, although usually in lower concentrations than in the serum. As might be expected, spermatozoa from such individuals exhibit a reduced ability to penetrate cervical mucus *in vitro*. Sperm samples collected from the spouse's cervix are generally immotile and no spermatozoa are found in the uterine cavity.

A second sperm antibody type that causes immobilization of the spermatozoon requires the presence of complement, and appears to correlate more closely with infertility than the sperm-agglutinating antibodies. Scanning electron microscopy often reveals sperm membrane damage induced by these cytotoxic antibodies and complement (Fig. 6.21).

Immunofluorescence can be used to localize antibodies on the surface of spermatozoa. This technique may detect antigens beneath the cell membrane of the spermatozoa since the sperm cells are usually exposed to a fixative, which alters the cell membrane and exposes internal antigens. Usually sperm smears are incubated with the serum to be tested and then with a fluorescent-labelled antibody such as anti-IgG or IgM. Antibodies specific for the acrosomal, equatorial, postacrosomal or neck region, and for the principal portion of the tail or the tail tip have been observed (Fig. 6.22). Individual sera will cause a characteristic staining pattern, which

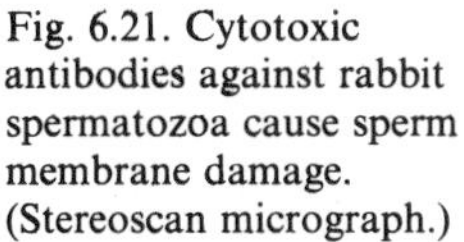

Fig. 6.21. Cytotoxic antibodies against rabbit spermatozoa cause sperm membrane damage. (Stereoscan micrograph.)

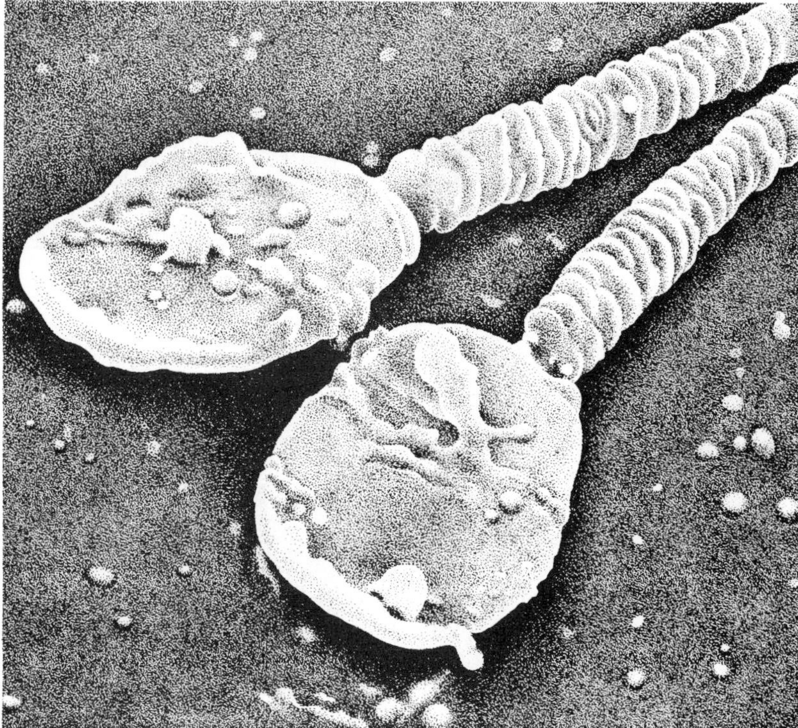

indicates that there are different antigens at different sites on the surface of the spermatozoon. Even when spermatozoa from different donors are exposed to the same sera, the same patterns will be seen, evidence that such sites are specific and show little variation between individuals. There are also many similarities between antibody types in males and females. Both sexes can develop the three types of antibodies – sperm-agglutinating, sperm-immobilizing (cytotoxic) and those revealed by immunofluorescence.

For circulating antibodies to be a factor in fertility regulation, they must come into contact with spermatozoa. None of the antibodies discussed here appears to affect spermatogenesis directly, owing to the competence of the blood–testis barrier. However, antisperm antibodies can be found in seminal plasma or in cervical mucus, and locally produced secretory IgA can develop in both males and females. In females the cervical glandular epithelium is considered capable of producing a local immune response. Antibodies in semen could be expected to affect sperm motility, viability, or the ability to fertilize the egg. Antibodies in cervical mucus or oviductal fluid would impede sperm transport.

Cellular immunity must also be considered in any discussion of immunological infertility, but it is logistically more difficult to envisage, since in normal women there are very few lymphocytes in the uterine lumen at

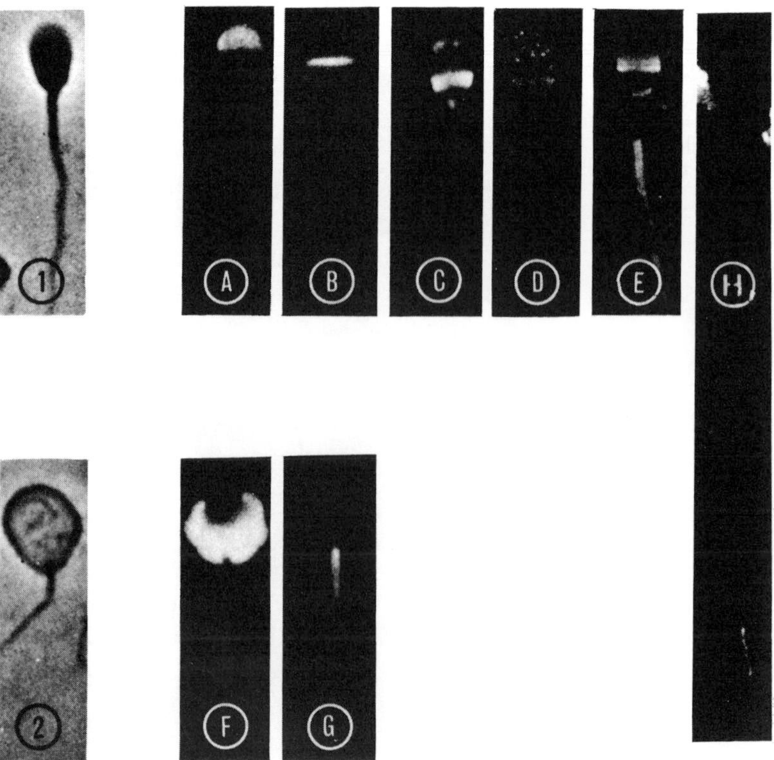

Fig. 6.22. Location of human sperm autoantibodies detected by indirect immunofluorescence. Regular sperm smears (1) permit detection of antiacrosomal 1 (A), antiequatorial (B), antipostacrosomal (C), antiacrosomal 2 (D), antitail, mainpiece (E), and antitip-of-tail (H) antibodies. Sperm smears pretreated with dithiothreitol and trypsin (2) show the distribution of antiprotamine (F) and antimidpiece (G) antibodies. (From K. S. K. Tung. *Clin. Exp. Immunol.* **20**, 93–104 (1975).)

the time of ovulation. In men, as we have mentioned, if the cellular immune system is involved, allergic orchitis might occur.

Studies on several fronts are under way to provide more information on how the immune system can affect fertility. For example, some workers are evaluating the use of immunosuppressive therapies in order to enhance the chances of pregnancy in patients with antibodies to spermatozoa. Other researchers are determining, by means of various functional tests, which antibodies may play a role in preventing cervical mucus penetration or ovum penetration. Such studies will provide not only information for treatment of infertility but possibilities for regulation of fertility by means of the immune system.

Immunoregulation of fertility

The possibility of using a unique antigen to control fertility is enticing. Development of any contraceptive vaccine requires the identification of physiologically active antigens specific to the human reproductive system that can be inactivated by antibody binding. Because of this, there has been much effort to identify and isolate antigens associated with reproductive processes. Three approaches are under consideration: (*a*) immunization against gamete-specific products, (*b*) immunization to substances produced by the conceptus, and (*c*) immunization to reproductive hormones. The prerequisites for a contraceptive vaccine for use in humans are as follows: (*a*) the vaccine must make use of pregnancy-specific or reproductive-tract-specific antigens that are not found on any other tissue in the body and produce a specific antibody response directed solely against the reproductive antigen; (*b*) the antigen preparation must be fairly easy to acquire and inexpensive; (*c*) the antigen should be immunogenic without the use of a noxious adjuvant; and (*d*) the contraceptive effect should be reversible.

The principle behind any contraceptive vaccine is that when reproductive-tract or pregnancy-specific antigens appear, the circulating antibodies combine with the antigens and inhibit the biological action of the target cell. Antibodies can impede reproductive functions by causing conformational changes in surface antigens, by blocking binding sites, by affecting motility or cellular interactions, or by inducing cell death (complement-mediated cytotoxicity). Since reproductive cycles and pregnancy are episodic, it is more likely that vaccines will be developed for use in women, even if the antigen is derived from men. Antibodies against spermatozoa in men could cause destruction of the germinal epithelium and thus irreversible sterility. To develop a vaccine, we must first identify the physiologically active proteins specific to the human reproductive system. In the past the approach has been to determine whether crude preparations were effective, and then in subsequent studies to work with purified antigens. However, since toxic substances or contaminants could cause the contraceptive effect, it is better to deal with purified substances from the start.

There are many theoretical problems associated with the development of contraceptive vaccines. A major concern, and one that is difficult to discount, even with extensive experimentation, is that the antigen must not cross-react with and cause damage to other tissues. Many antigenic determinants found in the reproductive tract are shared with normal tissue constituents and their products. For example, studies in the early 1960s indicated that the placenta and the basement membrane of the renal glomerulus have antigens in common; thus, immunization with placental tissue might cause renal problems. Non-identical antigens of a similar structure may allow a more generalized immune response.

Another concern is that we do not yet completely understand the maternal–fetal immune relationship. Therefore, there is always the possibility that incomplete effects of a contraceptive would result in abortion or teratogenesis (malformed but viable fetuses) or a higher incidence of neoplasms or intrauterine fetal growth retardation.

Because individual responses to antigens vary considerably, some individuals would show an intense immune response to the vaccine and others would not. In addition, contraceptive vaccines would probably require repeated immunizations; only certain viruses to which we are exposed during childhood provide life-long immunity. Researchers have frequently expressed the hope that the use of sperm antigens in a vaccine would allow sexual intercourse itself to provide a booster.

Below is a description of some of the antigen candidates for contraceptive vaccines.

Sperm antigens

Studies on sperm antigens and infertility have been conducted for many years. In 1932 Baskin injected wives with husbands' semen and rendered them sterile for about a year. Other investigators later demonstrated the development of circulating antisperm antibodies and allergic orchitis in men immunized with testicular homogenates. In both of these studies an adjuvant was used. As many as 30 to 40 different antigens are present in human semen, but which ones are targets for contraception is not known. Also, it is not easy to establish which antigens on ejaculated human spermatozoa are sperm-specific and which are acquired from the seminal plasma. An effective sperm antigen contraceptive vaccine in women must create high titres of antibodies against spermatozoa in the uterine and oviduct fluids. IgG levels are usually significantly lower in reproductive tract fluids than in serum; however, local production of IgA may be stimulated and antibodies secreted directly into the genital secretions. A promising antigen for an antisperm vaccine is lactic dehydrogenase C_4, an enzyme found only in spermatozoa and testicular cells. Recent studies in rabbits and baboons have demonstrated a significant reduction in fertility after immunization with this sperm antigen. Another is the sperm-specific enzyme acrosin, which is found in the sperm acrosome and is thought to play a role in egg penetration. Antibodies to such a sperm antigen could induce infertility in two ways, either by killing the spermatozoa that expressed the antigen or by blocking the function of the antigen.

Placental hormones

Human chorionic gonadotrophin (hCG). This glycoprotein is secreted by trophoblast cells within 6 to 8 days after fertilization and is thought to be important in the maintenance of pregnancy (see Book 2, Chapter 7, Second Edition). Human chorionic gonadotrophin stimulates the corpus luteum to continue producing oestrogen and progesterone and thus maintains the endometrium until the placenta takes over steroid hormone production 8 to 10 weeks later. Antibodies against hCG might interfere with maintenance of the endometrium or exert a cytotoxic effect on the developing blastocyst. Also, hCG may protect the developing conceptus from the maternal immune system by locally suppressing the function of maternal lymphocytes, or by coating antigens on the trophoblast cells and thus preventing immune recognition of foreign fetal tissue. Since hCG is produced very early in pregnancy and is found only in association with pregnancy (an exception is provided by certain tumours), it is an attractive target for fertility regulation. However, many of the pituitary hormones (e.g. thyroid-stimulating hormone (TSH), follicle-stimulating hormone (FSH), and luteinizing hormone (LH)) have a similar alpha subunit structure. To avoid immunological cross-reactivity, some investigators have used the beta subunit of hCG; if only the 30 amino acids at the carboxyl terminus of the beta chain are used, there is little cross-reaction even with LH, its closest chemical relative. Because the beta subunit is relatively non-immunogenic by itself (due to its small size), it must be coupled to larger 'carrier' molecules. A highly purified hCG subunit conjugated to tetanus toxoid (carrier) has been administered to women by Pran Talwar in India. In these experiments contraceptive effects were not the primary goal; the studies were designed to evaluate menstruation and ovulation, which continued normally. Contraceptive studies with hCG vaccines have been performed in baboons by Vernon Stevens in the United States and fertility has been greatly reduced. Studies on marmoset monkeys, conducted by John Hearn of Edinburgh, indicate that hCG-β suppresses fertility only while the antibody levels remain high. As levels fall, the female marmosets experience recurrent abortions interspersed with occasional live births. Whether a similar phenomenon might occur in some women remains an unanswered question.

Human placental lactogen. This hormone is synthesized by the trophoblast and secreted into the maternal circulation. It can be detected from week 5 of pregnancy, and the level rises progressively until term. During pregnancy this hormone stimulates breast growth and effects metabolic changes. Human placental lactogen (hPL) has a close physiological and chemical resemblance to human growth hormone. When antibodies to hPL were administered to pregnant rats, an intrauterine inflammatory response occurred and the pregnancies were terminated. Furthermore, the animals remained sterile for a long period after the passive immunization. Later studies showed that anti-hPL antibodies can also disrupt pregnancies in rabbits and baboons. However, because this vaccine causes abortion rather than blocks pregnancy before implantation, it offers a less attractive prospect than hCG vaccines.

Antigens of the trophoblast

Two other antigens of the placenta have been considered for fertility control. These are SP-1, a pregnancy-specific beta glycoprotein, and PP-5, a placental glycoprotein. The former is found in large quantities in pregnancy serum and can be made from the placentas of humans and rhesus macaques. It has a molecular weight of 90000 and is 28 per cent carbohydrate. Antiserum to human SP-1 produced in rabbits cross-reacts with placental extracts and pregnancy serum from monkeys and baboons. When ten cynomolgus macaques received injections of antiserum to human SP-1, eight animals aborted. Studies in baboons have demonstrated that abortion can occur as late as day 110 of pregnancy. This antibody is therefore an unlikely vaccine candidate. The other placental antigen, PP-5, is 10 per cent carbohydrate and has a molecular weight of 50000. Unfortunately, preliminary studies in monkeys show that immunization with a hapten coupled to human SP-1 also results in late abortion. Furthermore, PP-5 occurs in trace amounts, and it has been impossible to recover sufficient material to continue active work on it.

Zona pellucida antigens

The zona pellucida surrounding the egg contains unique antigens, and antibodies against the zona-specific proteins cause precipitins to form and thus prevent sperm penetration. Bonnie Dunbar and Paul Wasserman have recently made considerable progress in the identification and characterization of zona-specific antigens. It has been discovered that there are two principal proteins in the zona, both of which play important functional roles in sperm binding and egg penetration, and John Aitken has shown that reversible infertility can be produced in female marmoset monkeys immunized against zonal proteins from other species. Further progress in this field will be heavily dependent on genetic engineering techniques to produce an adequate supply of the antigen.

Stimulation of reproductive function by antibodies to ovarian steroids

Studies by Rex Scaramuzzi and his colleagues in Edinburgh and Australia, and Roger Land in Edinburgh (see Chapter 3) have led to an exciting new way of stimulating fertility in sheep. In ewes, quantitatively the most important steroid secreted into the ovarian vein at the time of oestrus is not oestradiol-17β but a biologically weak androgen, androstenedione (see Book 3, Chapter 5, Second Edition). In an effort to find out what this might be doing, antibodies were raised against it by coupling the steroid to a foreign protein and actively immunizing a group of ewes. The surprising result was that their ovulation rate was doubled. Subsequent studies showed that within certain defined limits of antibody titre, it was possible to achieve a highly significant increase in the number of lambs born – an effect that could also be produced over a shorter time span by passive immunization. The great advantage of this procedure is that, unlike ovarian stimulation with exogenous gonadotrophins, it never produces superovulations. Presumably the ovarian androstenedione is normally contributing, together with oestrogen, to the negative feedback inhibition of pituitary gonadotrophin secretion; by neutralizing this androgenic component, more gonadotrophin is produced, and so more follicles ovulate.

Inhibition of reproductive function by antibodies to gonadotrophin-releasing hormone

The hypothalamus regulates pituitary secretion of LH and FSH via the decapeptide gonadotrophin-releasing hormone (GnRH) secreted into the hypophyseal portal blood vessels. Studies in rats, rabbits, sheep, cattle and macaques indicate that successful immunization against the releasing hormone after coupling it to a large protein like albumin results in reduced levels of pituitary gonadotrophins in the blood, a phenomenon associated with involution of the testes and accessory glands in the male, and ovulatory failure in the female. Antibodies against GnRH thus are an important experimental tool for the evaluation of reproductive processes and could be invaluable for immunological castration of farm livestock or even domestic pets. Since men would lose their libido with the decline in testosterone secretion, this has no future as a contraceptive agent in men. Women, who probably are less dependent on their sex hormones for the maintenance of a normal libido, might accept active immunization. Some might welcome amenorrhoea as a by-product of contraception. As with other immunological approaches to fertility control that depend on the inhibition of a naturally occurring protein, an efficacy of less than 100 per cent would have to be acceptable.

Safety evaluations of vaccines

If a contraceptive vaccine is to be useful, its action must be reversible and safe. The criteria have to be much more stringent when the vaccine is designed to combat a condition in normal women that is not life-threatening. It is also important to know the type of immune response (humoral, cellular, or both) that would occur and whether the antibodies would cross-react with other tissues.

The immune system has been designed to seek out and kill foreign invaders. But the reproductive tract produces spermatozoa, ova and embryos that are themselves 'foreign', in the sense that the host has no natural tolerance to them. Therefore they need special protection from immunological attack, and we are only just beginning to discover the manifold ways in which this is achieved. When these protective mechanisms break down, sterility, or even death, may occur. We therefore need to exercise caution when evaluating the immunological consequences of a simple operation like vasectomy.

Immunological experience needs to be passed on from generation to generation. Although the newborn cannot in Lamarckian style inherit for life the acquired immunological experience of its mother, it can at least take out a long-term, interest-free loan by absorbing her antibodies following transfer across the placenta, or in the breast milk. This passive immunity suffices until such time as the neonate becomes immunologically competent to manufacture its own antibodies.

We are beginning to be able to put the immune system to use in order to regulate fertility. Not only can we stimulate fertility in sheep by immunizing them against their ovarian androgens, but we can also inhibit fertility in a whole range of laboratory animals by immunizing them against hypothalamic, pituitary, gonadal or placental hormones, or against specific proteins made by the gametes themselves. Unfortunately, the inherent variability of the immune response, and the need to use painful adjuvants in order to achieve high antibody titres, means that it will be some time before we have a reliable and acceptable form of immunological contraception for human use. But that should not stop us from trying to develop one.

Suggested further reading

Vasectomy increases the severity of diet-induced atherosclerosis in *Macaca fascicularis*. N. J. Alexander and T. B. Clarkson. *Science*, **201**, 538–41 (1978).

The development of immunoglobulin levels in man. M. Allansmith, B. H. McClellan, M. Butterworth and J. R. Maloney. *Journal of Pediatrics*, **72**, 276–90 (1968).

Placenta as an immunological barrier. A. E. Beer and J. O. Sio. *Biology of Reproduction*, **26**, 15–17 (1982).

Immunoreproduction: concepts of immunology. M. W. Chase. *Biology of Reproduction*, **6**, 335–57 (1972).

Differential effect of pregnancy or gestagens on humoral and cell-mediated immunity. N. Fabris, L. Piantanelli and M. Muzzioli. *Clinical and Experimental Immunology*, **28**, 306–14 (1977).

Immunobiology of mammalian reproduction. A. E. Beer and R. E. Billingham. *Advances in Immunology*, **14**, 1–84 (1976).

The Immunobiology of Mammalian Reproduction. A. E. Beer and R. E. Billingham. Prentice-Hall; Englewood Cliffs, New Jersey (1976).

Immunology. J. A. Bellanti. Saunders; Philadelphia (1971).

Immunological Influence on Human Fertility. B. Boettcher. Academic Press; Sydney (1977).

Immunological Response of the Female Reproductive Tract. B. Cinader and A. de Weck. Scriptor; Copenhagen (1976).

Development of Vaccines for Fertility Regulation. M. J. K. Harper, W. R. Jones, R. M. Y. Ing, E. R. Hobbin, T. Hjort, C. R. Austin, C. A. Shivers, V. C. Stevens, H. Bohn and K. S. K. Tung. Scriptor; Copenhagen (1976).

Bioregulators of Reproduction. G. Jagiello and H. J. Vogel. Academic Press; New York (1981).

Immunology of Human Reproduction. J. S. Scott and W. R. Jones. Academic Press; London (1976).

Reproduction and Antibody Response. S. Shulman. CRC Press; Cleveland (1975).

Acknowledgement

The work described in this chapter, Publication No. 1236 of the Oregon Regional Primate Research Center, was supported by NIH Grant RR-00163 and NICHD Contract NO1-HD-8-2827.

7

Reproductive senescence

*C. E. ADAMS**

Senescence, the process or condition of growing old – characteristically progressive and irreversible – affects reproduction just as it does other vital functions. Indeed, Edward Schneider and Richard Blandau in their prefaces to two recent books on ageing have each remarked on the value of the female reproductive system and of the gametes, the eggs and spermatozoa, as ideal models for the exploration of other ageing systems.

Gerontology, the study of ageing, has very recent origins, barely 3 decades ago. It is hardly surprising, therefore, that there are still many gaps in our knowledge of ageing, both in a general context and specifically in relation to reproduction. Nevertheless, within the last decade significant advances have occurred, notably as a result of the development of sensitive hormone assay procedures.

In human populations clinical interest in ageing is considerable and likely to increase as the proportion of the elderly grows. Since 1900 the average life expectancy in England has increased from 47 to 71 years and, according to WHO, by the year 2000, it is expected to rise to between 75 and 80 in developed countries and 65 and 70 years in developing countries.

Apart from what we know of man, satisfactory data on the relation between lifespan and reproduction are available for only a few species, mostly short-lived laboratory animals and certain wild animals kept in zoos. Records of farm animals tend to be misleading as only very select individuals are usually retained for breeding purposes, the majority being

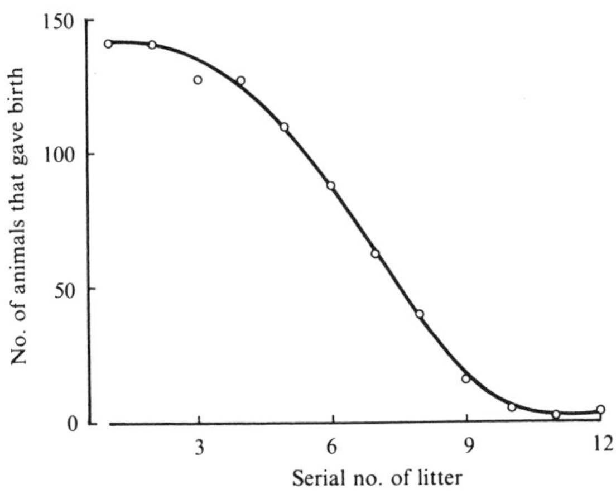

Fig. 7.1. Decline in number of fertile rats with birth of successive litters. (From D. L. Ingram, A. Mandl and S. Zuckermann. *J. Endocr.* **17**, 280–5, fig. 1 (1958).)

* Dr C. E. Adams died on 29 March 1984.

killed for a variety of reasons. Even with no selection, the age-related decline in fertility tends to be very rapid, as shown in Fig. 7.1. Investigation of the mechanisms of ageing is plagued by the expense of keeping experimental animals for long periods during which many inevitably die or have to be culled without contributing to the results.

The course of senescence

In litter-bearing (polytocous) animals, such as the mouse, litter size is related to litter order (parity) and maternal age; there is an initial rise, a plateau period and subsequent decline. This is illustrated diagrammatically

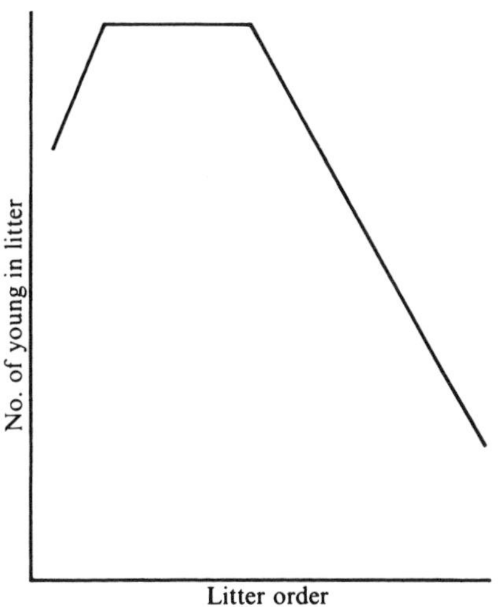

Fig. 7.2. Diagram showing the general relation between litter size and litter order in individual female mice. The reproductive life history consists of an initial rise, a plateau period and a period of linear decline. (From J. D. Biggers, C. A. Finn and A. McLaren. *J. Reprod. Fert.* **3**, 313–30, text-fig. 3 (1962).)

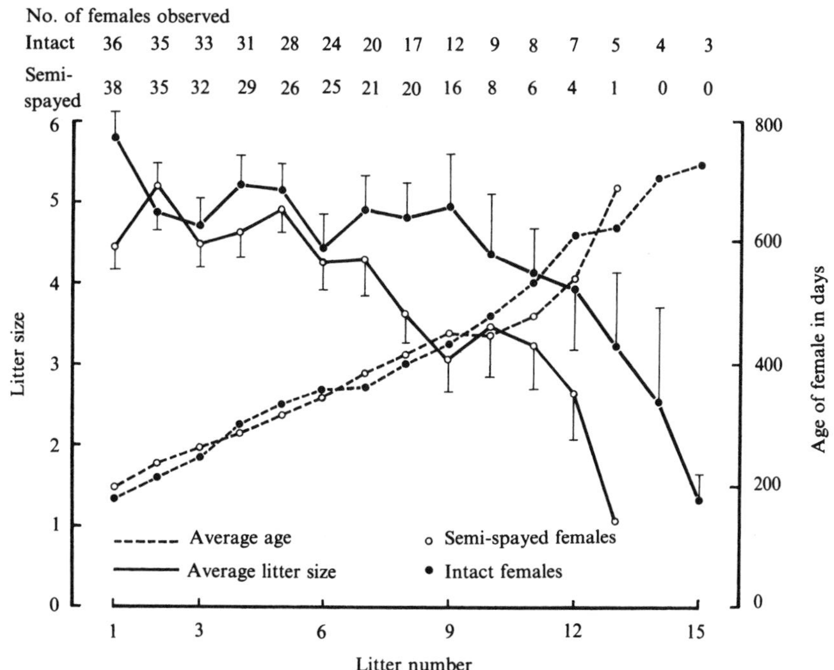

Fig. 7.3. Litter size according to parity and age in intact and semi-spayed Mongolian gerbils. (From M. L. Norris and C. E. Adams. *Lab. Anim.* **16**, 146–50, text-fig. 1 (1982).)

in Fig. 7.2. In practice, the decline is rarely abrupt, but occupies a significant part of the animal's reproductive lifespan; actual examples appear in Figs. 7.3 and 7.4, whilst others (e.g. Fig. 7.5) have been provided by Marie-Claire Levasseur and Charles Thibault. During the decline phase, a decreasing litter size is merely one manifestation of a general decay in reproductive function; other signs include changes in the frequency and regularity of oestrous or menstrual cycles and, in monotocous species particularly, a decline in the chances of becoming pregnant. It has been estimated that, in women over 35, one-third of conceptions will end in spontaneous abortion, probably very early, mostly due to a rise in the frequency of trisomic abnormalities (see Book 2, Chapter 5, Second Edition). The incidence of stillbirths and abnormal births is also greatly increased (Table 7.1).

There are several possible explanations for reproductive decline. Helen King in 1916 attributed it to a reduction in the number of eggs shed, and also questioned whether 'abnormal ova are more frequent in old animals than in young ones and so help diminish fertility in later life'. Yet another possibility was put forward by Sidney Asdell and his co-workers of Cornell, namely that the accessory organs may wear out so that eggs cannot be fertilized or embryos carried to term. In about 1960 Esther Jones and Peter Krohn in Birmingham established, from painstaking studies on

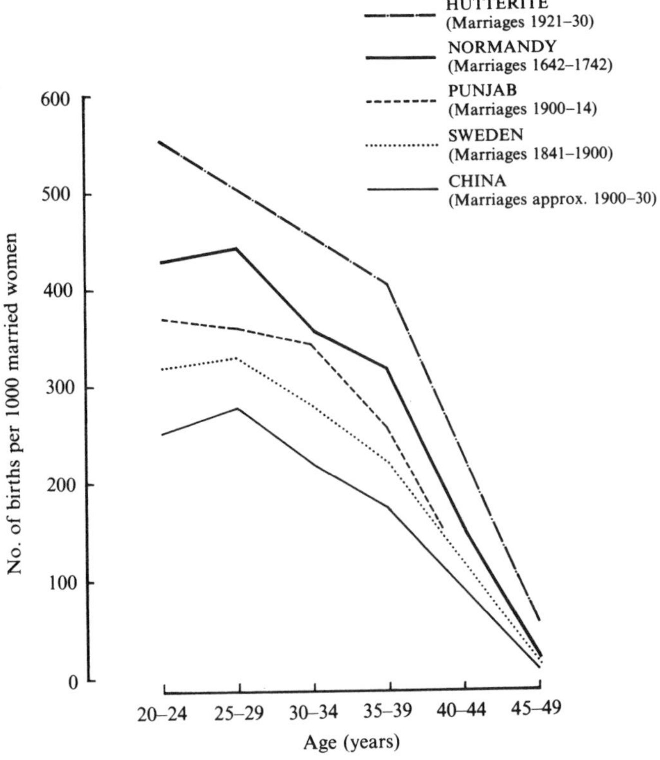

Fig. 7.4. Age-specific marital fertility for selected populations. (From R. H. Gray. *J. Biosoc. Sci.* Suppl. **6**, 97–115 (1979).)

oocyte populations in the mouse ovary, that the loss of fertility in this species was not due to exhaustion of the oocyte store. In man, on the other hand, oocytes are nearly totally depleted at the time of the menopause. Jones and Krohn concluded that the reason for decline in fertility in the mouse was 'more likely to be found in the hormonal control of the ovary or in the uterine environment'.

More recent observations based on reciprocal embryo transfer between young and old laboratory animals indicate that the decline is due to increasing embryonic mortality rather than to a fall-off in ovulation rate, and it appears that the ageing uterus is primarily to blame. Experiments on mice, hamsters and rabbits have proved beyond doubt that the uterus of aged mothers cannot support the development of embryos from young donors. In two of these species, rabbit and hamster, it appears that embryos obtained from old animals are also suspect, for very few were able

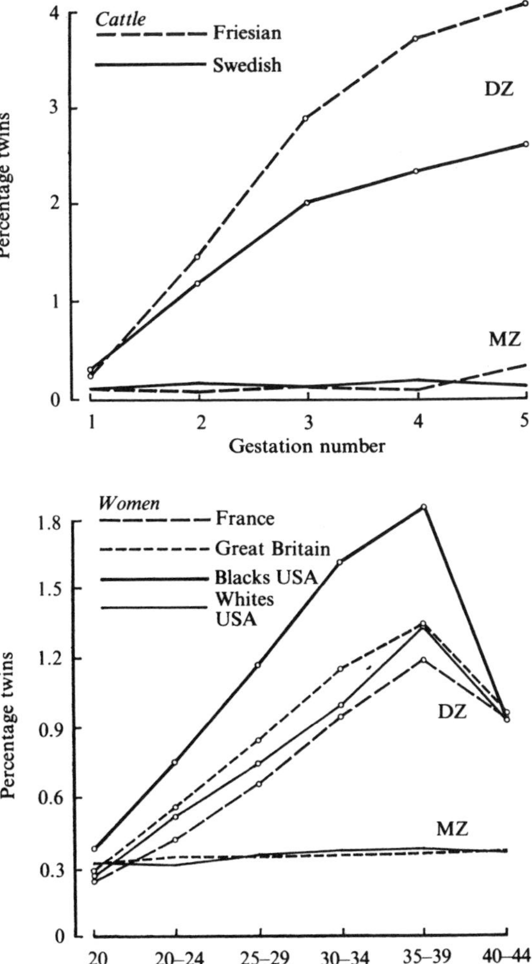

Fig. 7.5. The incidence of twin births in cattle and women (DZ, dizygotic; MZ, monozygotic). (From M.-C. Levasseur and C. Thibault. *De la Puberté à la Sénescence. La Fécondité chez l'Homme et les autres Mammifères.* Masson; Paris (1980).)

to develop in the healthy uteri of young recipients (see Table 7.2). However, mouse embryos developed equally well in young recipients irrespective of the donor's age, a result subsequently confirmed by Roger Gosden.

Discussion of the ovarian contribution has been renewed following the work of Alan Henderson and Robert Edwards of Cambridge on chiasma frequency, which led to their 'production-line' hypothesis. They propose that eggs are ovulated in the order in which they start upon the first meiotic division, and that a later start carries a greater risk of chromosomal

Table 7.1. *Proportion of stillbirths in all births by age of mother: France (1974), England and Wales (1974). Anomalies at birth: France, 1963–9*

Age of mother (years)	Stillbirths/1000 births: France	Stillbirths/1000 births: England and Wales	Abnormal births/1000 live births: France, 1963–9
13–19	14.0	10.8	16.1
20–24	12.2	9.5	14.4
25–29	12.2	9.5	17.7
30–34	15.3	10.5	19.1
35–39	22.7	16.9	29.1
40–44	35.1	24.5	—
45–49	58.5	45.2	48.7
All ages	14.0	10.3	

(From H. Leridon, *J. Biosoc. Sci.* Suppl. **6**, 59–74 (1979).)

Sources:

Office of Population Censuses and Surveys (1978). Marriage and Divorce Statistics, and Birth Statistics, England and Wales 1975. HMSO; London.

Institut National de la Santé et de la Recherche Medicale (INSERM) (1978) Malformations Congénitales. Risques Prénatales (Enquête Prospective, 1963–9). INSERM; Paris.

Institut National de la Statistique et des Études Economiques (INSEE) (1977) La Situation Démographique en 1974. Collection D50. INSEE; Paris.

Table 7.2. *Survival of fertilized eggs to birth after transfer between animals of different ages*

	Eggs surviving to term (%): Rabbit	Hamster	Mouse
Young to old	1.5	8.3	14.0
Young to young	50.0	49.2	48.0
Old to young	12.5	4.5	54.0

(From C. E. Adams, *J. Reprod. Fert.* Suppl. **12**, 1–16 (1970).)

abnormality. This leaves the abnormal forms to be ovulated in the ageing animal when the oocyte store is becoming depleted.

In ageing rabbits fertilization is less reliable than in young animals, where it is common to find 100 per cent of the eggs fertilized. Additionally, in eggs that are penetrated by spermatozoa a two- to seven-fold increase in cleavage failure has been observed (Fig. 7.6). An increased incidence of abnormal eggs has also been noted in the golden hamster, mouse and rat. We may well ask to what extent these results are the expressions of inherent defects in the egg or embryo, attributable to ageing changes in the maternal environment. By transferring unfertilized eggs obtained from old animals shortly after ovulation to young recipients it has been found that the primary defect seems to lie within the egg.

In the female mammal no single component of the genital system appears able to escape the consequences of senescence, with certain organs or functions being affected earlier or more severely than others. Species and individual variations further complicate matters.

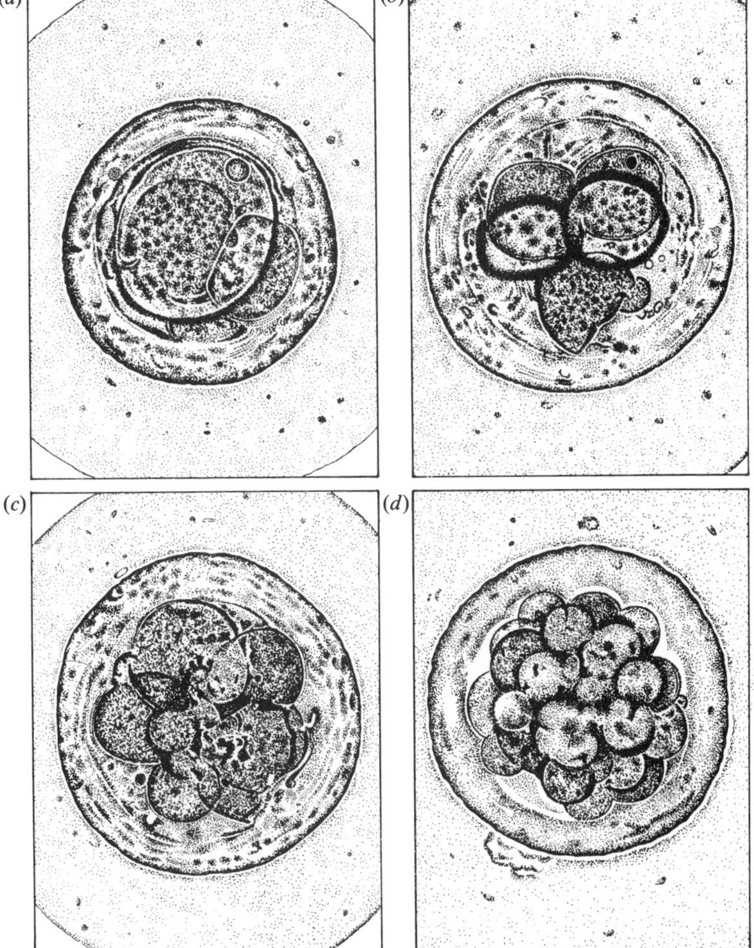

Fig. 7.6. (*a*), (*b*) and (*c*). Examples of eggs showing arrested cleavage, recovered from the ovidict of an old rabbit 60 h post-coitum. (*d*) A normal egg (morula stage) recovered from the oviduct of a young rabbit 60 h post-coitum. (From C. E. Adams. *J. Reprod. Fert.* Suppl. **12**, 1–16, plate 1 (1970).)

No discussion of reproductive senescence would be complete without reference to the hypothalamic–pituitary system, because of its key role in reproduction and the growing realization of its susceptibility to ageing, whether inherent or indirect via feedback mechanisms.

Owing to fundamental differences in gametogenesis and steroid hormone synthesis, the effects of ageing in the male are less obvious than in the female. Nevertheless, there is a progressive decline of sexual activity with age in men (Fig. 7.7), though the influence of diminished health rather than old age on sexual parameters can confound the issue. Basal testosterone and oestradiol levels did not differ between younger (24–37 years) and older (60–88 years) men, but the response to 2 days of hCG stimulation decreased significantly with age (Fig. 7.8), irrespective of the level of sexual

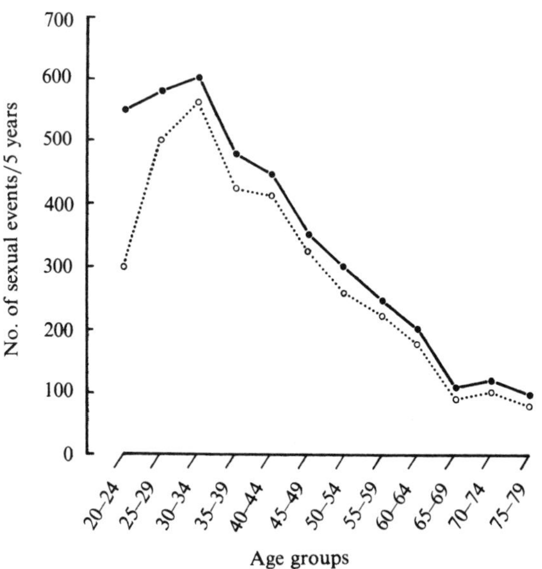

Fig. 7.7. Decline of sexual activity with age in men. Solid line, total sexual activity, including masturbation, nocturnal emissions, etc.; dotted line, coital activity. (From C. E. Martin. In *Handbook of Sexology*, p. 815, fig. 1. Ed. J. Money and J. Mousaph. Elsevier/North Holland Biomedical Press; Amsterdam (1977).)

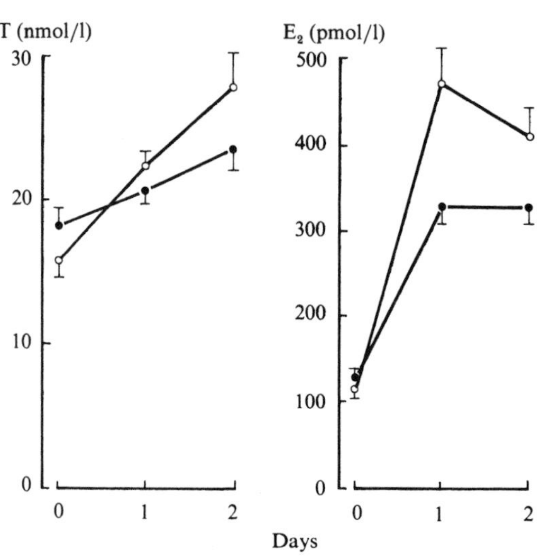

Fig. 7.8. Responses to hCG stimulation in 11 young fathers (○) and 23 grandfathers (●). Points are the mean ±SE. On each day three blood samples were taken from each subject. T, testosterone; E_2, oestradiol. (From E. Nieschlag, V. Lammers, C. W. Freischeim, K. Langer and E. J. Wickings. *J. Clin. Endocr. Metab.* **55**, 676–81, fig. 8 (1982).)

activity. Recent work on rats, in which plasma free testosterone declines after 12 months of age, indicates that the levels of aromatizing enzymes in the brain and plasma testosterone are correlated.

Unlike oogenesis, spermatogenesis continues from puberty throughout life, and both the 'quality' of the ejaculate and fertility appear to be quite well maintained (Table 7.3). In man there are several well-authenticated reports of the aged, even a 94-year-old, begetting offspring. A recent study by Ebo Nieschlag and his co-workers, using zona-free hamster eggs, showed that the fertilizing capacity of spermatozoa from elderly men remained unaltered.

Observations on male mice suggest that their infertility – most are sterile by 24 months of age – may be due to physical incapacity rather than to reduction in the fertilizing capacity of their spermatozoa. In the bull there

Table 7.3. *Seminal parameters in young fathers and grandfathers*

	Young men (n = 20)	Old men (n = 22)	P
Ejaculate volume (ml)	4.0 ± 1.7	3.2 ± 1.9	NS
Sperm density (millions/ml)	78 ± 51	120 ± 101	<0.05
Sperm motility (%)	68 ± 14	50 ± 19	<0.0005
Normally formed sperm (%)	52 ± 13	48 ± 9	NS
Fructose (mg/ml)	3.2 ± 1.1	1.7 ± 1.1	<0.0005
pH	7.28 ± 0.09	7.25 ± 0.15	NS
Hamster-oocyte-penetration test (% penetrated ova)	54 ± 20	54 ± 18*	NS

Values are the mean ± SD. NS, not significant. * n = 16.
(From E. Nieschlag, V. Lammers, C. W. Freischeim, K. Langer and E. J. Wickings, *J. Clin. Endocr. Metab.* **55**, 676–81 (1982).)

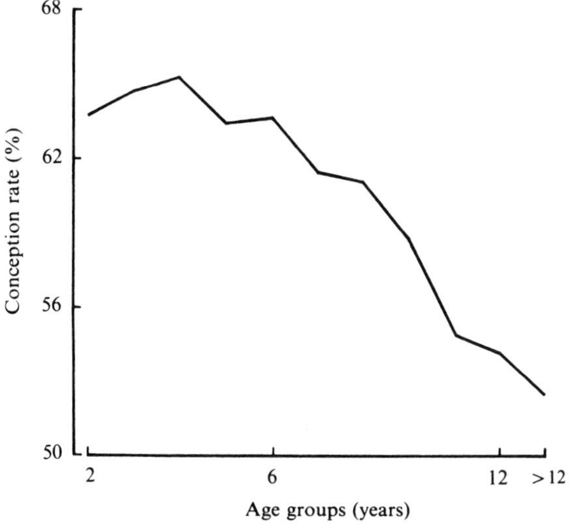

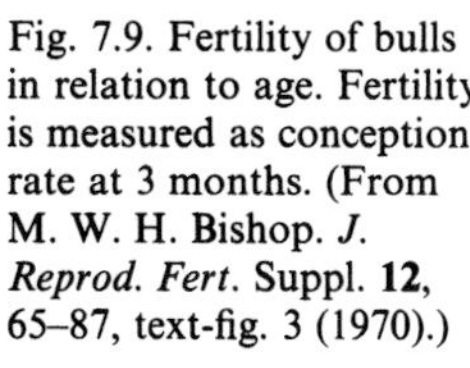

Fig. 7.9. Fertility of bulls in relation to age. Fertility is measured as conception rate at 3 months. (From M. W. H. Bishop. *J. Reprod. Fert.* Suppl. **12**, 65–87, text-fig. 3 (1970).)

is evidence that fertility falls with age, even among highly fertile animals retained for AI; here the physical aspect can be excluded. Over 60000 artificial inseminations during 1946–49 showed greatest fertility at 3–4 years of age and then a slow decline of about 1 per cent per year (Fig. 7.9).

Gametes

Oocytes

There is a growing awareness of the potentially harmful effects of gamete ageing, particularly of the human egg. Ageing may occur at several different stages during the interval from oogenesis to ovulation, or after ovulation if sperm penetration is delayed.

With few exceptions, the process of oogenesis in a mammal is already complete at the time of, or very soon after, birth, and the stock of oocytes will not be added to later in life. Consequently, in women an oocyte may have been present in the ovary for 40 years or more before it is ovulated. One of the best known and most tragic conditions attributable to ageing is Down's syndrome, or mongolism, due to trisomy of chromosome 21 (further details are given in Book 2, Chapter 5, Second Edition). The incidence of Down's syndrome is of the order of 1 in 2000–2500 for mothers under 30 years, 1 in 1200 between 30 and 34 years and 1 in 300 at 35–39 years, with a progressively steeper rise after this age. The pattern appears to be the same irrespective of racial, geographic or socio-economic factors. The incidence of Down's syndrome as related to age is shown in Fig. 7.10, which also shows the high incidence of spontaneous abortions. David Carr has estimated that as much as 40 per cent of early fetal wastage may be due to chromosome abnormalities. The exact cause of this

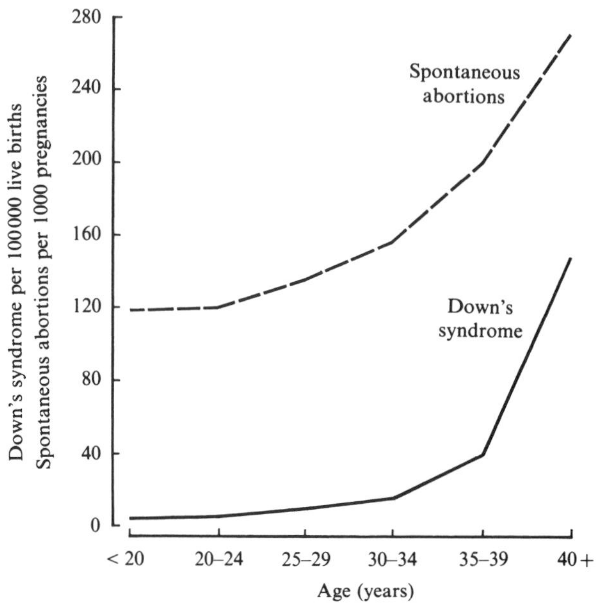

Fig. 7.10. The age-specific incidence of spontaneous abortions and Down's syndrome. (From R. H. Gray. *J. Biosoc. Sci.* Suppl. **6**, 97–115, fig. 3 (1979).)

age-dependent increase is not known. Before the 'production-line' hypothesis (p. 214), the favoured explanation was that ageing during the long dictyate stage might progressively impair the proper placement and attachment of chromosomes on the spindle during the first meiotic division. Another possibility is that oocytes ovulated later in life are subjected to environmental influences (e.g. radiation, drugs) not experienced to the same degree by those ovulated early in life. The process of ovulation may also be impaired in the aged ovary so that the egg tends to deteriorate before fertilization. To some extent the increase in the incidence of abnormal offspring may be peculiar to man. It has been suggested that, in polytocous species (litter-bearers), intrauterine competition may eliminate abnormal conceptuses, but this remains an open question.

Ageing of the oocyte within the Graffian follicle. In cyclic female rats ovulation can be delayed by 1 or 2 days by treatment with appropriate drugs, such as pentobarbital sodium. Such delayed ovulation may be associated with the production of abnormal zygotes; in rats a threefold increase in polyspermy was observed, and increased embryonic mortality. Jongbloet has suggested that there may be a link between the incidence of anencephaly and oocyte maturity in women, and Hideo Nishimura has speculated from his observations that delayed ovulation in women may be responsible for the occurrence of some severe malformations. The picture is complicated by the fact that many other links in the chain of processes preceding ovulation may be disrupted, and this might contribute to embryonic mortality. In aged females there is evidence of pituitary LH deficiency, which could delay the process of ovulation, causing the oocyte to become overripe.

Ageing of the oocyte after ovulation. In general the fertilizable life of mammalian eggs is very short and of the order of a few hours. Even so, it tends to exceed the period within which fertilization will lead to the birth of a normal healthy offspring. In most species the existence of an oestrous state ensures that spermatozoa are already waiting at the site of fertilization in the oviduct before the eggs are shed. An obvious and most important exception to this rule is the human species in which sexual intercourse may occur at any time in the menstrual cycle. Consequently, if the frequency of intercourse is low – and as it decreases with age (see Fig. 7.7) – aged oocytes are more likely to be fertilized in elderly couples.

When rats are mated or artificially inseminated late in relation to ovulation, there is appreciably more abnormal fertilization, mainly polyspermy. Normally mammalian eggs exhibit a 'block to polyspermy' which restricts the entry of extra spermatozoa; this involves a change in either the zona pellucida, as in the rat (and in man and farm animals), or at the surface of the vitellus, as in the rabbit. The effectiveness of this blockade may be gauged from the fact that with normal time relations the incidence of polyspermic fertilization rarely exceeds 1–2 per cent, but when mating is delayed the frequency can rise tenfold, owing, it is thought, to deterioration in the ageing eggs. Polyspermic fertilization with one extra spermatozoon leads to triploidy, which is now recognized as a common cause of miscarriage early in human pregnancy.

Spermatozoa

Spermatozoa may be subjected to ageing both before ejaculation in the male tract and after ejaculation in the female tract, as well as under artificial conditions *in vitro*. When rabbit testicular spermatozoa are experimentally prevented from passing into the epididymis by ligation of the vasa efferentia, their fertilizing capacity is maintained for 38 days, and their motility for as long as 60 days. Similarly, in both the rat and guinea-pig motility lasts somewhat longer than the ability to fertilize. Decreasing fertility is accompanied by an increase in the frequency of intrauterine mortality and fetal abnormalities. This problem has also been studied in the rabbit by testing the effect of extending the interval between semen collections; even with intervals as long as 40 days there were no significant differences in semen quality, fertilization rate or embryonic mortality that could be ascribed to collection interval. This is probably because spermatozoa are continuously eliminated from the male tract. Moreover, aged spermatozoa fare badly when in competition with fresh spermatozoa, as demonstrated in poultry and rabbits, despite the fact that they may be fertile when used alone.

Ageing of spermatozoa in the female tract may be achieved by carrying out inseminations at progressively longer intervals before ovulation. The rabbit is ideal for such studies because ovulation is non-spontaneous and can easily be induced by injection of human chorionic gonadotrophin or

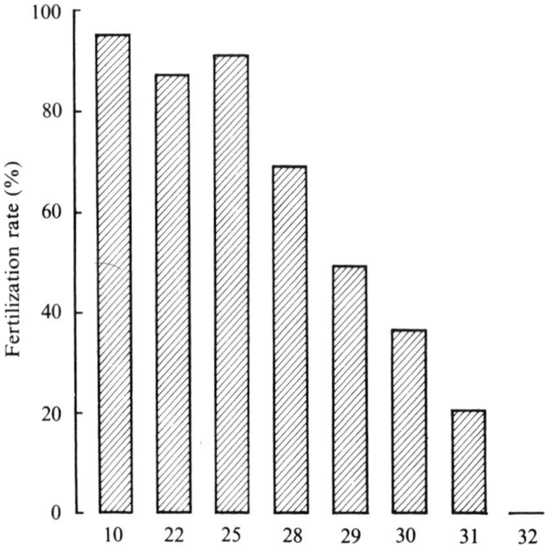

Fig. 7.11. Effect of ageing of spermatozoa on fertilization rate in the rabbit. (From J. M. Tesh. *J. Reprod. Fert.* **20**, 299–306, text-fig. 1 (1969).)

by allowing mating with a sterile male. Rabbit spermatozoa retain their fertilizing capacity for 25 h in the female tract, but thereafter a decline sets in so that after 31 h fertility is lost (see Fig. 7.11). Under similar conditions the fertile lifespan of deep-frozen spermatozoa is much reduced: they have lost their fertilizing capacity by 20 h following insemination. It appears that the main effect is on the fertilization process, in contrast to the increased incidence of embryonic mortality following experimentally induced ageing in the male reproductive tract.

With the availability of techniques such as electron microscopy, the study of sperm ageing is no longer confined to the whole cell; observations on individual organelles have focused attention on changes in the cell membranes and the acrosome.

Storage of spermatozoa in vitro. Much work has been done on the storage of spermatozoa *in vitro*, particularly of bull spermatozoa on account of the bull's economic importance in national artificial insemination schemes. The effect of length of storage and of temperature on fertility is shown in Fig. 7.12. After an unexpected initial increase attributed to the selective elimination of defective spermatozoa, fertility declines at a rate depending upon the storage temperature; in the opinion of George Salisbury and Robert Hart in Urbana, this is due to increasing embryonic mortality caused by changes in the genetic information carried by the spermatozoa.

Capacitation. During their passage through the female tract, spermatozoa of some species, if not all, undergo a process termed 'capacitation', which prepares them for penetration of the egg membranes. Once spermatozoa have become capacitated, their lifespan may be quite limited. On the other hand, the 'fertilizing lifespan' of rabbit spermatozoa is extended if they are held in the uterus of a female rat. Under these conditions the

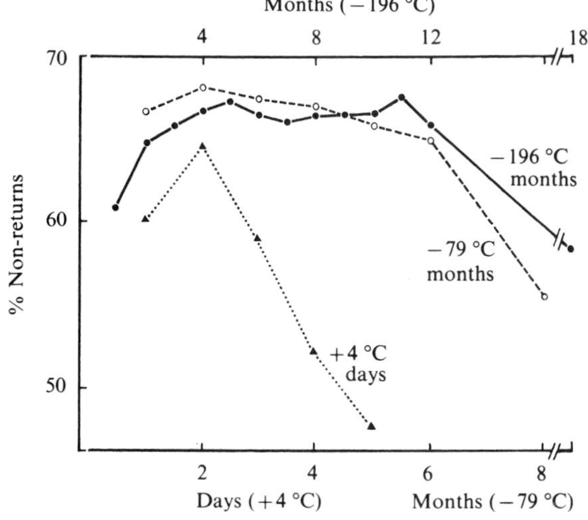

Fig. 7.12. The effect of duration of storage and of temperature on fertility of bull spermatozoa. (From G. W. Salisbury and R. G. Hart. *Biol. Reprod.* Suppl. **2**, 1–13, fig. 1 (1970).)

capacitation process does not reach completion and fertilizing capacity is retained for the whole motile life of the spermatozoon, which may amount to 50–55 h.

Ovary

According to Alex Comfort, senescence of the gonad regularly precedes or accompanies senescence of its owner in a number of phyla, to the extent that declining reproductive capacity provides a valuable expression of senescence. Thung noted that the progression of the life cycle from juvenile development through adult function to senile involution is more pronounced in the ovary than in any other organ. Estimates of the hormonal lifespan of the ovaries in relation to total lifespan are presented in Table 7.4.

Table 7.4. *Hormonal lifespan of the ovaries in relation to total lifespan*

		Percentage of total lifespan spent in			
Species	Average expected lifespan	Immature non-cyclic phase	Regular cyclic phase	Irregular cyclic phase	Senile non-cyclic phase
Human	75 years	16	44	7	33
Rhesus monkey	30 years	7	67	16	10
Mouse (*C57BL*)	26 months	8	58	19	15
Mouse (*CBA*)	28 months	7	32	14	47
Rat	33 months	8	43	39	10

(From E. C. Jones. In *Psychosomatics in Peri-menopause*, pp. 13–39. Ed. A. A. Harper and H. Musaph. MTP Press; Lancaster (1979).)

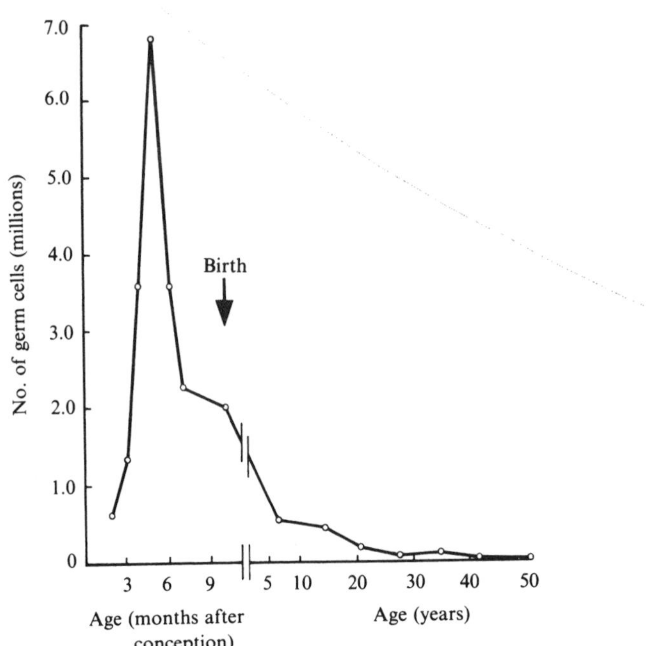

Fig. 7.13. Fluctuations in total population of germ cells in the human ovary during reproductive life. (From T. G. Baker. *Amer. J. Obstet. Gynec.* **110**, 746–61, text-fig. 1 (1971).)

Number of oocytes

Much of what we know about oogenesis and the number of oocytes present in the mammalian ovary was derived in the 1960s in man, monkey, guinea-pig, mouse, rat, ferret, golden hamster, rabbit, cow, sheep and pig, but for the majority of animals the facts are still unknown. At birth the stock of oocytes is measured in many thousands, e.g. in man estimates range from 200000 to more than half a million, but most are destined to degenerate by a process known as atresia; the reduction in oocyte numbers from conception to 45 years is shown in Fig. 7.13 (see also Book 1, Chapter 2, Second Edition). Only a few will actually be ovulated and of these, even in the most fecund species, no more than about 150 are likely to be fertilized and develop to term. The study of Esther Jones and Peter Krohn in Birmingham on mice already referred to (p. 212) will serve to illustrate the relationship between age, numbers of oocytes and fertility. They found strain variations in the age of development reached at birth, in the total numbers of oocytes present and in the rates at which the oocytes are lost (Fig. 7.14). Only in one of their strains, identified as *CBA*, did the ovary become totally depleted of oocytes long before death; at the same age, 437 days, strain-*A* ovaries contained an average of about 365 oocytes, strain-*R* III ovaries 540, and *CBA* × *A* hybrid ovaries about 865, suggesting the existence of hybrid vigour. Reproduction had no significant effect on the rate at which the total number of oocytes declined. The *CBA* strain lost 28 per cent of the existing oocytes every 28 days, whilst in the other strains the loss was only half this amount.

Ovarian response to pituitary hormones

In some animals, such as the cow, the ovary is already capable of responding at birth to injections of follicle-stimulating hormone, FSH, with distinct follicular enlargement, though ovulation rarely occurs; in others, no follicular response may be evoked for weeks or even months. This variation depends upon the stage of differentiation of the ovary, which

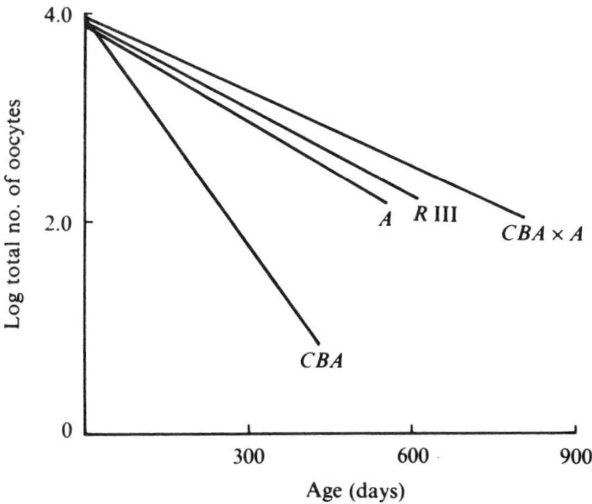

Fig. 7.14. Combined regression lines showing the relationship between age and total numbers of oocytes in various strains of mice. (From E. C. Jones and P. L. Krohn. *J. Endocr.* **21**, 469–95, text-fig. 9 (1961).)

in some species is well advanced at birth. The appearance of vesicular (antral) follicles containing liquor folliculi distinguishes the onset of response. In the rabbit this stage is reached at about 11 weeks of age when the first ovulations can be induced, normally well before the animal is capable of maintaining pregnancy.

Ovarian response rises to a maximum at 5–6 months in the rabbit and then levels off before declining significantly with age. Likewise, the natural ovulation rate declines by 30 to 40 per cent in aged does, presumably reflecting the decline in the number of large follicles, as occurs in the mouse (Fig. 7.15); in women the rate drops from a maximum of 231 Graafian follicles at 16 years to about 70 between 18 and 31 years, and to lower levels between 39 and 44 years. The ageing human ovary exhibits an increasing resistance to gonadotrophins.

Endocrine function

Apart from its gametogenic role, the ovary is a most important endocrine organ, being responsible for the secretion of oestrogen and progesterone, but this function waits upon the ripening of follicles, ovulation and the development of corpora lutea. During the period of lowered fertility there is no direct evidence, except in women, that the secretion of either of these hormones is significantly reduced. Indeed, very recent work on rats aged 4–22 months with normal ovarian cycles revealed little if any effect of ageing on progesterone secretion, and, by inference, on alterations in the neuroendocrine control of pituitary hormone secretion. In general, reproduction seems to end well before endocrine function becomes impaired. The latest ages at which young corpora lutea and large follicles were found in virgin mice are given in Table 7.5.

Thung, in his review on ageing changes in the ovary, presented as evidence for the decreasing efficiency of ovulation and corpus luteum formation in ageing mice the frequency with which follicles undergo luteinization without previous ovulation (Table 7.6). Studies on ageing

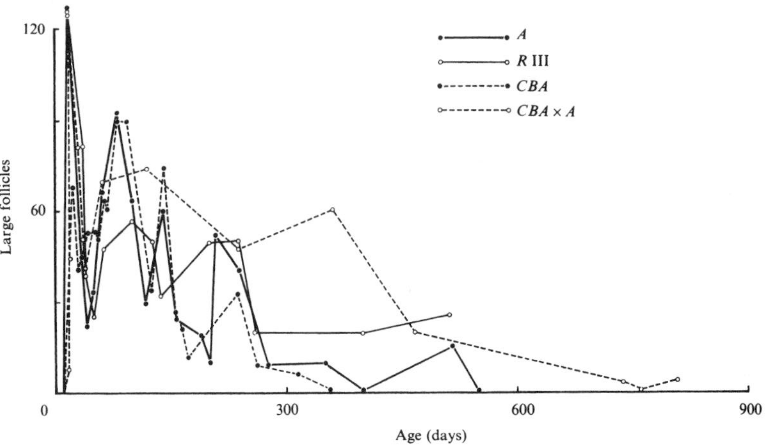

Fig. 7.15. Numbers of large follicles in ovaries of virgin mice. (From E. C. Jones and P. L. Krohn. *J. Endocr.* **21**, 469–95, text-fig. 11 (1961).)

mice suggest that the amount of circulating LH becomes inadequate or disappears altogether, and it may be that this is attributable to ageing of the hypothalamus.

Genital tract

In ageing female mammals declining ovarian function is shown by loss of weight of the organ and a reduction in numbers of both follicles and corpora lutea (see Table 7.7). An extreme condition is found in women at the menopause when ovarian function fails, and subsequently the genital tract atrophies.

Ageing may also lead to changes of a pathological nature, some of which are gross and immediately recognizable, whilst others are at the cellular level and require microscopic study and expert interpretation. In the first category a good example is the condition known as 'cystic ovary', the incidence of which is clearly dependent on age in rodents, and possibly in

Table 7.5. *Latest ages at which young corpora lutea and large follicles were found in virgin mice*

Strain of mouse	Age at which young corpora lutea present (days)	Large follicles present (days)
CBA	320	360
A	525	561
R III	100	589
CBA × *A*	480	852

(From E. C. Jones and P. L. Krohn, *J. Endocr.* **21**, 469–95 (1961).)

Table 7.6. *The number of corpora lutea and the percentage of luteinized atretic follicles (corpora lutea enclosing an oocyte) in serially sectioned mouse ovaries at different ages*

Age in months	No. of ovaries sectioned	Average total no. of corpora lutea per ovary	Luteinized atretic follicles (%)
Strain *C57BL*			
16–17	12	7.9	5
18–21	12	5.6	28
Hybrids $F_1(O_{20} \times DBA_f)$			
14–19	17	14.7	2
20–25	32	4.8	36
26–30	28	1.0	32

(From P. J. Thung. In *Structural Aspects of Ageing*, Ch. 9, pp. 109–42. Ed. G. H. Bourne and E. M. H. Wilson. Pitman Medical; London (1961).)

other orders, too, though comparative data are lacking. Information on the incidence of the condition in rats and hamsters is given in Table 7.7, and in the Mongolian gerbil *Meriones unguiculatus* in Fig. 7.16. Among females aged 200–400 days, only 5 per cent had cystic ovaries, whereas in those examined at 600–900 days 73 per cent were found to have cysts, and

Table 7.7. *Comparison of ovarian changes between adult and aged rats and hamsters*

	Rat		Hamster	
	Adult (3 months)	Aged (22–32 months)	Adult (3 months)	Aged (21 months)
No. of animals/group	8	28	5.0	5.0
Ovarian weight (mg/ovary/animal)	49±2.0*	36±4.0	Not reported	Not reported
No. of follicles/ovary	29±3.0	9±0.9	42.8	12.4
Mean no. of CL†/ovary	15±1.4	8±1.0	4.8	3.0
Animals with cystic follicles (%)	0	43	0	20.0
Animals with 'wheel' cells in ovaries (%)	0	75	Not reported	Not reported

* Mean±SD. † Corpora lutea.
(From A. P. Labhsetwar, *J. Reprod. Fert.* Suppl. **12**, 99–117 (1970).)

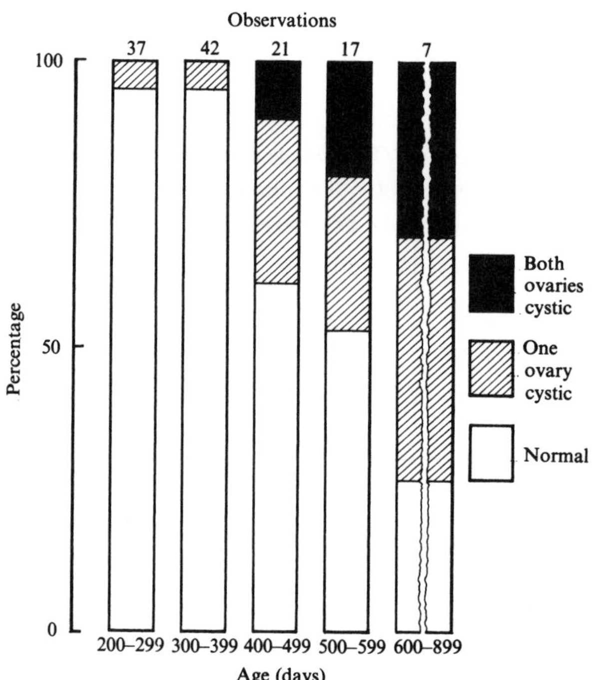

Fig. 7.16. The incidence of cystic ovaries according to age in the Mongolian gerbil. (The last column is split because it covers a much longer time interval than the others.) (From M. L. Norris and C. E. Adams. *Lab. Anim.* **6**, 337–42 (1972).)

in an increasing proportion both ovaries were affected. Sometimes the cysts assumed massive proportions, accounting for up to one-sixth of the animal's weight. The ovarian tissue becomes stretched to a thin film on the surface of the cyst and conception fails. However, when only one ovary is affected the other may continue to function normally, showing that the effect is quite local.

The second kind of pathological change seen with ageing in the ovary includes certain forms of cellular degeneration (ceroid and amyloid), the accumulation of so-called 'wheel' cells and recognizably sterile follicles, and the formation of testis-like tubules and ingrowths of the germinal epithelium. Ovaries from ageing multiparous mice were found to have fewer histological abnormalities than those from ageing virgin mice. Ovarian cancers are relatively uncommon; in ageing rats and mice the incidence of primary ovarian tumours is only about 0.03 per cent and 0.2 per cent, respectively.

In women, the genital tract often becomes diseased to the extent that major surgery is required. A Dutch survey based on more than 4000 women aged between 42 and 62 years revealed that one-sixth had undergone surgery of the uterus or ovaries. In the rabbit, too, the uterus is particularly prone to disease, and tumour formation is common. It has been found that up to 90 per cent of does have diseased uteri, especially between 4 and 6 years of age (see Table 7.8). Before tumours form, the endometrium exhibits a cystic appearance (cystic glandular hyperplasia) and fails to respond to progesterone, as shown in Fig. 7.17. The condition causes sterility, for embryos cannot survive in such an environment. Since cystic hyperplasia also occurs commonly in women before the menopause, it may well contribute to the declining fertility that characterizes this phase, but we have little specific information on the effects of uterine ageing in monotocous species, including man.

According to Marcus Bishop the degenerative changes that occur in the testis 'appear to be related to no other obvious factor than advancing age itself'. A recent report (1980) based on 238 bulls culled from the AI service by the Milk Marketing Board shows that the average score of testicular

Table 7.8. *Incidence of diseased uteri relative to age in the rabbit*

		Condition of uterus					
		'Normal'		One horn diseased		Both horns diseased	
Doe's age (years)	No. of does	No.	%	No.	%	No.	%
4.0–4.9	22	10	45	7	32	5	23
5.0–6.0	29	2	7	1	3	26	90

(From C. E. Adams, *J. Reprod. Fert.* Suppl. **12**, 1–16 (1979).)

lesions (tubular atrophy, fibrosis, interstitial cell hypoplasia, etc.) increases in a linear manner with age (Fig. 7.18). However, differences were noted between Friesian bulls culled for infertility, with high testicular lesion scores at all ages, and those slaughtered for other reasons, in which the score increased gradually with age, reaching the high levels seen in the infertile group only in extreme old age.

Hemi-ovariectomy

When one ovary is removed in polytocous animals, the remaining one compensates by shedding roughly the same number of eggs as the two had done before, and fertility, at least initially, does not suffer. This fact, established nearly two centuries ago by John Hunter in two pigs, has been found true for all species so far examined, including the opossum, golden hamster, rat, mouse, rabbit and Mongolian gerbil (see Fig. 7.3). Moreover, if all but a fragment of ovarian tissue is removed, that fragment can hypertrophy to approximate the size of two normal ovaries, providing the operation is performed early in life.

Information as to whether hemi-ovariectomy in women hastens the onset of the menopause is scarce, though it is known that in mice the rate

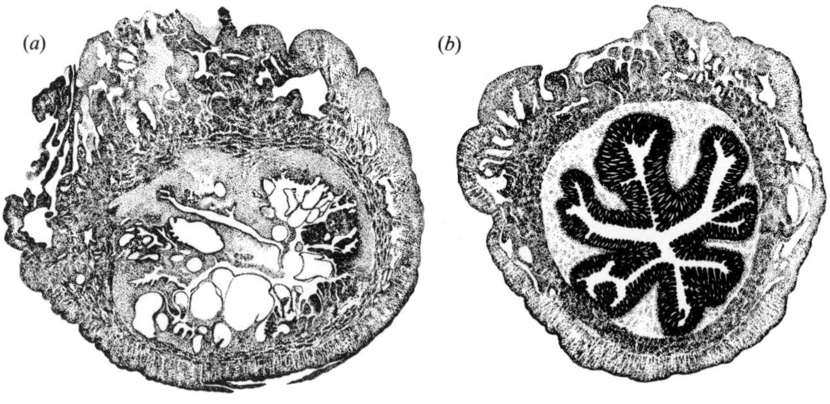

Fig. 7.17. (*a*) Section of uterus, 60 h post-coitum, from a multiparous rabbit doe aged 50 months. Note extensive cystic enlargement (hyperplasia) and lack of endometrial growth. (*b*) Section of uterus, 96 h after first mating, from a doe aged 7 months. Note extensive endometrial growth. (From C. E. Adams. *Preimplantation Stages of Pregnancy*, pp. 345–73, figs. 11 and 10, respectively. Ed. G. E. W. Wolstenholme and M. O'Connor. Churchill; London (1965).)

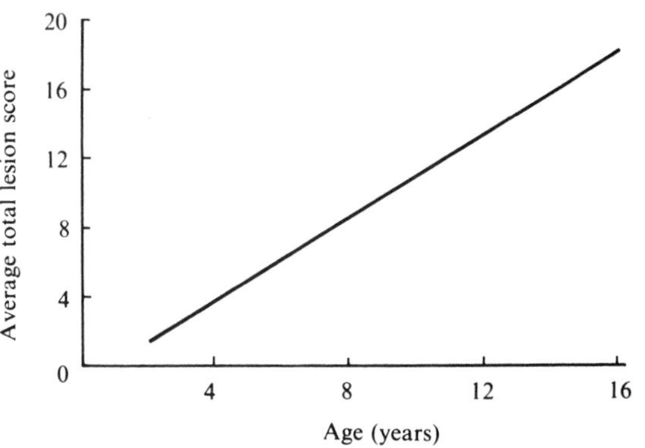

Fig. 7.18. The relationship between testicular pathology and age in the bull. (From Report of the Breeding and Production Division, Milk Marketing Board, Thames Ditton, Surrey, No. 30. 1979/80.)

of loss of oocytes in a single ovary is substantially greater than would be expected to occur in one ovary of a normal animal. Removal of one ovary in the mouse does not disturb the normal oestrous cycle, but the cycles cease earlier in life than in control animals (Fig. 7.19).

Experiments on mice designed to test the long-term effects of ovariectomy on reproductive performance have produced some unexpected and potentially far-reaching results. Although at first the performance of the one-ovary mice equalled that of their controls, after about the sixth pregnancy they ceased to reproduce. In the mouse, unlike the pig, migration of eggs from one uterine horn to the other is a very rare event; thus after hemi-ovariectomy pregnancy is normally confined to one horn which therefore carries double its normal load. The uterus appears to become prematurely exhausted and although the exact point of breakdown has not been defined, impairment of the vascular supply is suspected.

When the experiment was repeated on rats a similar overall result was

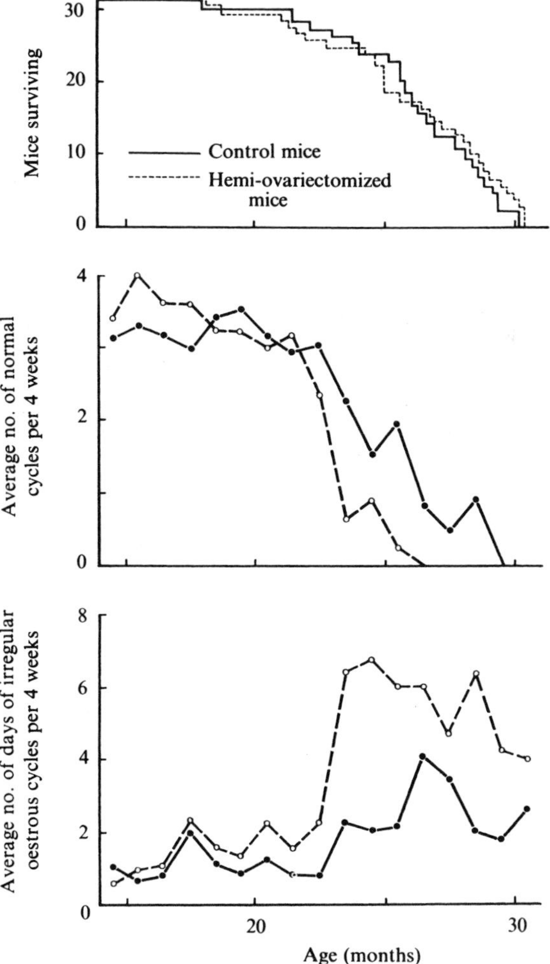

Fig. 7.19. Ageing changes as exemplified in the oestrous cycles of normal and unilaterally ovariectomized F_1 ($O_{20} \times DBA_f$) mice. The animals were allowed to live their maximum lifespan. The average number of normal and irregular cycles per animal were calculated over 4-week intervals. (From P. J. Thung. In *Structural Aspects of Ageing*, pp. 109–42, fig. 10. Ed. G. H. Bourne. Pitman Medical Press; London (1961).)

obtained though the pattern of reproduction was slightly different. In this case, too, the sixth pregnancy represented a dividing line, but now reproduction fell to a low level rather than ceased altogether. In fact, the mean age of the hemi-ovariectomized rats when they produced their final litter equalled that of their controls. However, the one-ovary rats did produce significantly fewer litters, reflecting a longer interval between successive pregnancies. This feature is also reminiscent of the effect of ageing on reproduction. When representatives of the two groups were examined in their sixth pregnancy, the number of ovulations was found to be similar, but the level of embryonic mortality was nearly twice as high in the one-ovary group as in the controls.

Climacteric

Probably more is known of the terminal stages of reproductive life in man than in any other mammal; a recent WHO report on this subject contains over 380 references. A distinguishing feature, apparently shared by very few other species, is the menopause, when the menses cease, proclaiming the final decay of ovarian function. At this point the oocyte stock becomes exhausted and the production of ovarian oestrogen ends, causing the degenerative changes associated with the menopause. Clinically, the climacteric (that phase in the ageing process of women marking the transition from the reproductive stage of life to the non-reproductive stage) is especially significant because it usually involves endocrine, somatic and psychic changes, as illustrated in Fig. 7.20.

Although we have long known that gonadotrophin levels rise at the time of the menopause as a result of the absence of ovarian steroid feedback, it is only within the last year or two that we have begun to realize that the chemical nature of the gonadotrophins secreted by the pituitary also

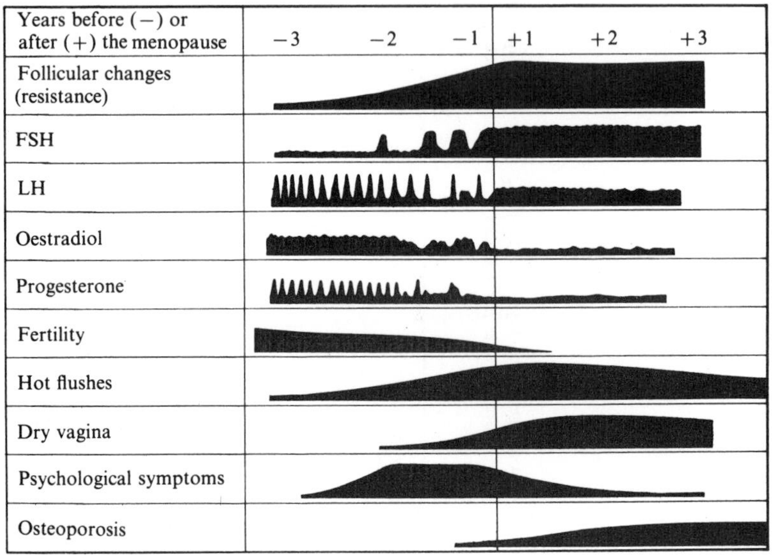

Years before (−) or after (+) the menopause	−3	−2	−1	+1	+2	+3
Follicular changes (resistance)						
FSH						
LH						
Oestradiol						
Progesterone						
Fertility						
Hot flushes						
Dry vagina						
Psychological symptoms						
Osteoporosis						

Fig. 7.20. Schematic representation of some clinical, biological and endocrinological features of the peri- and post-menopausal phases. (From *Research on the Menopause*. Report of a WHO Scientific Group, 670. WHO; Geneva (1981).)

changes. Leif Wide in Uppsala and Bruce Hobson in Edinburgh have shown that postmenopausal women secrete a type of FSH, identical to that found in intact or castrated men, that is more acidic, has a longer half-life and is more biologically active than that secreted by premenopausal

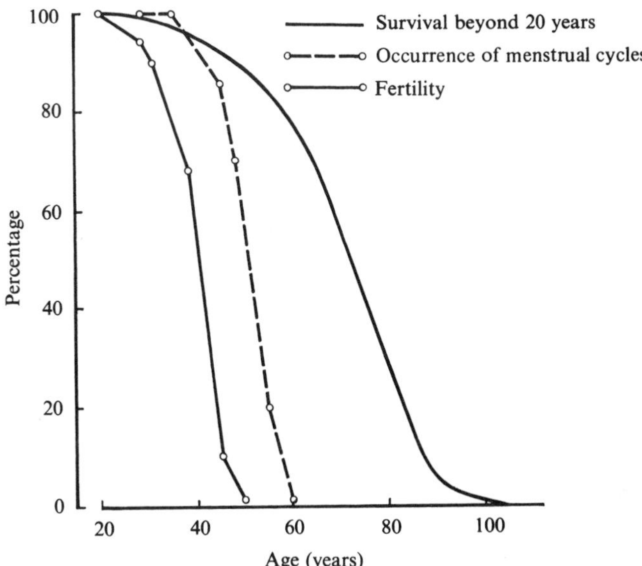

Fig. 7.21. Life expectancy and fertility in women. (From M.-C.-Levasseur and C. Thibault. *De la Puberté à la Sénescence. La Fécondité chez l'Homme et les autres Mammifères*. Masson; Paris (1989).)

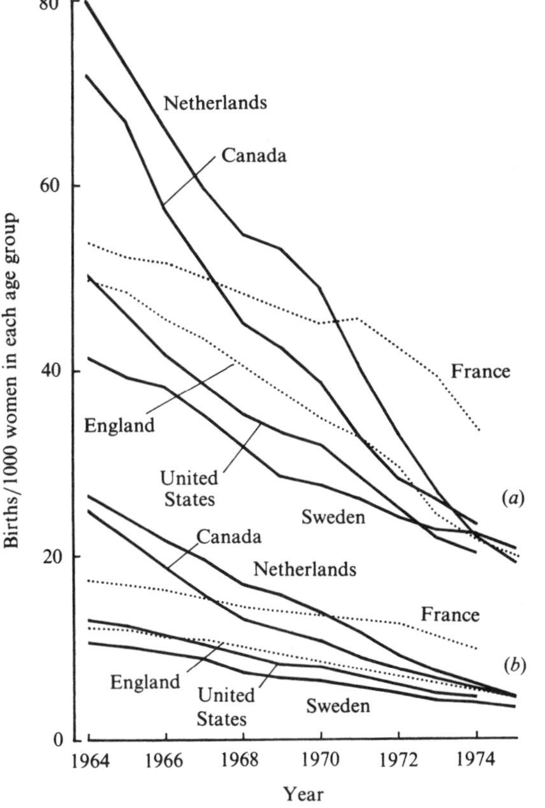

Fig. 7.22. Age-specific fertility rates, 1964–65: (*a*) age 35–39, (*b*) 40–44. (From H. Leridon. *J. Biosoc. Sci.* Suppl. **6**, 59–74 (1979).)

women. This is apparently due to a direct effect of ovarian oestrogen on the pituitary, altering the sialic acid content of the secreted FSH.

Contrary to the belief that the menopause is occurring later in life, the contemporary view is that the average age at menopause, now considered to be 50–51 years (range 42–60 years), has probably not altered substantially over the centuries, though the age at menarche almost certainly has. Many women menstruate regularly till the age of 54 or 55 years, but conception after the age of 50 is rare. The relationship between life expectancy, the menopause and fertility is shown in Fig. 7.21.

In England and Wales in 1967, the incidence of pregnancy in women over 44 years of age was only 0.1 per cent. Even among the Hutterites, a strictly religious sect living in Alberta and South Dakota, who do not practise contraception, childbearing ceases at about 49 years of age. Undoubtedly, the occurrence of pregnancy in ageing women is now limited by socio-economic pressures and the use of contraceptives. Data from the United States show that in 1910 11 per cent of women aged between 45 and 49 years of age had children under 5 years old, whereas in 1950 (before introduction of the Pill) this figure had shrunk to 5 per cent. The proportion of births in women aged 35 and over among all births in France and England and Wales fell from 13.8 per cent and 11.3 per cent in 1964 to 7.7 per cent and 5.9 per cent, respectively, in 1975 (Fig. 7.22). Nowadays, therefore, the post-reproductive period of life often spans three decades or more.

Ageing affects many functions and nowhere is this more evident than in reproduction and the processes of development. It is a change that is not only inevitable but also continuous and irreversible, going on throughout life; as a popular aphorism puts it, we start to age as soon as we are born. Even this is not the whole truth, for with the occurrence of pre-ovulatory ageing, it would seem that we may begin to age before we begin!

Suggested further reading

Reproductive functions in young fathers and grandfathers. E. Nieschlag, U. Lammers, C. W. Freischem, K. Langer and E. J. Wickings. *Journal of Clinical Endocrinology and Metabolism*, **55**, 676–87 (1982).

Aging in the hypothalamic–hypophyseal ovarian axis in the rat. P. Ascheim. In *Hypothalamus, Pituitary and Aging*, pp. 376–418. Ed. A. V. Everitt and J. A. Burgess. Thomas; Springfield, Illinois (1976).

The menopause (a wall-chart). R. J. Beard, R. H. Gray and H. S. Jacobs. *Research in Reproduction*, **10**, no. 6 (1978).

Qualitative difference in follicle-stimulating hormone activity in the pituitaries of young women compared to that of men and elderly women. L. Wide and B. M. Hobson. *Journal of Clinical Endocrinology*, **56**, 371–5 (1983).

The post fertile life of non-human primates and other mammals. E. C. Jones. In *Psychosomatics in Peri-menopause*, pp. 13–39. Ed. A. A. Harper and H. Musaph. MTP Press; Lancaster (1979).

Gamete aging and its consequences. G. W. Salisbury and R. G. Hart. *Biology of Reproduction*, Supplement 2, 1–13 (1979).

Aging of the reproductive system. G. B. Talbert. In *Handbook of the Biology of Aging*, pp. 318–56. Ed. C. E. Finch and L. Hayflick. Van Norstrand Reinhold; New York (1977).

Aging Gametes Their Biology and Pathology. Ed. R. J. Blandau. Kager; Basle (1975).

The Biology of Senescence, 3rd ed. A. Comfort. Churchill Livingstone; Edinburgh (1979).

The menopause. *Clinical Obstetrics and Gynaecology*, vol. 4. Ed. R. B. Greenblatt and J. Studd. W. B. Saunders; London (1977).

De la Puberté à Sénescence. La Fécondité chez l'Homme et les autres Mammifères. M.-C. Levasseur and C. Thibault. Masson; Paris (1980).

Fertility in middle age. Ed. A. S. Parkes, M. A. Herbertson and J. Cole. *Journal of Biosocial Science*, Supplement no. 6, 223 pp. Galton Foundation; Cambridge (1979).

Aging reproductive system. E. L. Schneider. *Aging*, vol. 4. Raven Press; New York (1978).

Research on the Menopause: Report of a WHO Scientific Group. World Health Organization Technical Report Series 670, 120 pp. WHO; Geneva (1981).

INDEX

Numbers in italics indicate pages on which data appear in Figures; those printed **bold** *are for pages with relevant Tables*